Energy and Geometry

An Introduction to Deformed Special Relativity

World Scientific Series in Contemporary Chemical Physics – Vol. 22

Energy and Geometry

An Introduction to Deformed Special Relativity

Fabio Cardone

National Research Council, CNR, Roma, Italy

Roberto Mignani

Università di Roma "Roma Tre", Italy

World Scientific

NEW JERSEY • LONDON • SINGAPORE • BEIJING • SHANGHAI • HONG KONG • TAIPEI • CHENNAI

Published by

World Scientific Publishing Co. Pte. Ltd.

5 Toh Tuck Link, Singapore 596224

USA office: 27 Warren Street, Suite 401-402, Hackensack, NJ 07601

UK office: 57 Shelton Street, Covent Garden, London WC2H 9HE

Library of Congress Cataloging-in-Publication Data
Cardone, Fabio.
   Energy and geometry : an introduction to deformed special relativity / Fabio Cardone,
Roberto Mignani
      p. cm. -- (World Scientific series in contemporary chemical physics ; v. 22)
   Includes bibliographical references and index.
   ISBN-13 978-981-238-728-8 (alk. paper)
   ISBN-10 981-238-728-5 (alk. paper)
   1. Special relativity (Physics) 2. Generalized spaces. I. Mignani, Roberto. II. Title.
   III. Series.

QC173.65.C35 2004
530.11--dc22

                                                                    2004049708

British Library Cataloguing-in-Publication Data
A catalogue record for this book is available from the British Library.

Typeset by Stallion Press
Email: enquiries@stallionpress.com

Printed in Singapore by World Scientific Printers (S) Pte Ltd

To our parents

Maria Pia, Mario   Nietta[a], Pietro

---

[a]After the completion of the book, the mother of R.M. passed in May 2003. R.M. dedicates this book to her memory.

Preface

In 1955 the Italian mathematician Bruno Finzi, in his contribution to the book *"Fifty Years of Relativity"*,[b] stated his *"Principle of Solidarity"* (PS),[c] that sounds *"It's (indeed) necessary to consider space-time TO BE SOLIDLY CONNECTED with the physical phenomena occurring in it, so that its features and its very nature do change with the features and the nature of those. In this way not only (as in classical and special-relativistic physics) space-time properties affect phenomena, but reciprocally phenomena do affect space-time properties. One thus recognizes in such an appealing 'Principle of Solidarity' between phenomena and space-time that characteristic of mutual dependence between entities, which is peculiar to modern science."* Moreover, referring to a generical N-dimensional space: *"It can, a priori, be pseudoeuclidean, riemannian, non-riemannian. But —* he wonders *— how is indeed the space-time where physical phenomena take place? Pseudoeuclidean, riemannan, non-riemannian, according to their nature, as requested by the principle of solidarity between space-time and phenomena occurring in it."*

Of course, Finzi's main purpose was to apply such a principle to Einstein's Theory of General Relativity, namely to the class of gravitational phenomena. However, its formulation is as general as possible, so to apply in principle to all the known physical interactions. Therefore, Finzi's PS is at the very ground of any attempt at geometrizing physics, i.e. describing physical forces in terms of the geometrical structure of space-time.

---

[b]B. Finzi: "Relatività Generale e Teorie Unitarie", in *"Cinquant'anni di Relatività"*, ed. M. Pantaleo (Giunti, Firenze, Italy, 1955), p. 194.

[c]It is quite difficult to express in English in a simple way the Italian words *"solidarietà"* and *"solidale"*, used by Finzi to mean the feedback between space-time and interactions. A possible way to render them is to use *"solidarity"* and *"solidly connected"*, respectively — at the price of partially loosing the common root of the Italian words — with the warning that what Finzi really means is that the very structure of space-time is determined by the physical phenomena which do take place in it.

Such a project (pioneered by Einstein himself) revealed itself unsuccessful even when only two interactions were known, the electromagnetic and the gravitational one. It was fully abandoned starting from the middle of the 20th century, due to the discovery of the two nuclear interactions, the weak and the strong one (apart from recent attempts based on string theory).

Actually, the very euclidean geometry is not only the first historical example of self-consistent and complete, axiomatic system of mathematics, but does represent the first theory of physical reality (as stressed for instance by Penrose). It describes in a quantitative way, in mathematical language, the relations among measured physical entities — distances, in this case — and therefore of the physical space in which phenomena occur.

However, the measurement of distances depends on the motion of the body which actually performs the measurement. Such a dependence is indeed not on the *kind* of motion, but rather on the *energy* needed to let the body move, and on the *interaction* providing such energy. The measurement of time needs as well a periodic motion with constant frequency, and therefore it too depends on the energy and on the interaction.

This simple example shows how *energy does play a fundamental role in determining the very geometrical structure of space-time* (in analogy with the General-Relativistic case, where — as is well known — the energy-momentum tensor is the source of the gravitational field).

By starting from such considerations, a possible way to implement Finzi's principle for *all* fundamental interactions is provided by the formalism of *Deformed Special Relativity* (*DSR*) developed in the last decade. It is based on a *deformation* of the Minkowski space, namely a space-time endowed with a metric whose coefficients just depend on energy. Such an energy-dependent metric does assume a *dynamical role*, thus providing a geometrical description of the fundamental interaction considered and implementing the feedback between space-time structure and physical interactions which is just the content and the heritage of Finzi's principle.

Aim of this book is to acquaint the reader with the basic concepts and methods of DSR. Part I and II contain the physical and mathematical foundations of the theory. In Part III, we discuss the physical experiments whose analysis allows one to derive the explicit form of the phenomenological metrics for all four fundamental interactions, and some of the physical applications of the DSR formalism. The connection of DSR with the problem of the breakdown of local Lorentz invariance is considered in Part IV.

The natural link between DSR and a generalized, five-dimensional Kaluza-Klein scheme is also outlined.

We are indebted of course to many colleagues and friends. First of all, it is both a duty and a pleasure to express our special gratitude to Myron Evans, who was so kind to invite us to write this book for the series "*Advances in Chemical Physics*" of which he is the Editor in Chief. Part of the content of the book was the object of a one-month lecture course delivered by F.C. in spring 2002 at the Mathematical Department of Messina University, on invitation by Enzo Ciancio and Liliana Restuccia, and we are pleased to thank them warmly. We are very grateful to the co-authors of all our papers, in particular to Umberto Bartocci, Mauro Francaviglia, Mario Gaspero, Alessio Marrani, Vladislav S. Olkhovsky, Eliano Pessa. Thanks are also due to Carol O. Alley, Tito Arecchi, Gaetano Caricato, Sidney Coleman, Sheldon L. Glashow, Roman Jackiw, Maxim Yu. Khlopov, Rostislav V. Konoplich, Daniela Mugnai, Fabio Pistella and Anedio Ranfagni for kind interest and useful discussions. Needless to say, the responsibility for any errors or omission rests with the authors.

Fabio Cardone
Roberto Mignani
Roma, May 2003

Contents

III   Metric Description of Fundamental Interactions   **51**

PART I

GENERALIZING SPECIAL RELATIVITY

CHAPTER 1

AN AXIOMATIC VIEW TO SPECIAL RELATIVITY

1.1. Foundations of Special Relativity

Special Relativity (SR) is essentially grounded on the properties of space-time, *i.e.* isotropy of space and homogeneity of space and time (as a consequence of the equivalence of inertial frames) and on the Galilei principle of relativity.

The two basic postulates of SR in its axiomatic formulation are[1]:

1. *Space-time properties*: Space and time are homogeneous and space is isotropic.
2. *Principle of Relativity (PR)*: All physical laws must be covariant when passing from an inertial reference frame K to another frame K', moving with constant velocity relative to K.

In the second postulate it is clearly understood that, for a correct formulation of SR, it is necessary to specify the total class, C_T, of the physical phenomena to which the relativity principle applies. The importance of such a specification is easily seen if one thinks that, from an axiomatic viewpoint, the only difference between Galilean and Einsteinian relativities just consists in the choice of C_T (*i.e.* the class of mechanical phenomena in the former case, and of mechanical and electromagnetic phenomena in the latter).

Such kinds of phenomena are ruled by fundamental interactions derivable from scalar (algebraic) or vector (Euclidean) potentials. Mechanical phenomena, corresponding to the class C_T of Galilean relativity, are determined by a uniform and stationary gravitational field (fully described in terms of a Newtonian potential) whereas electromagnetic phenomena considered within SR are determined by the potential component of the electromagnetic interaction.

Depending on the explicit choice of C_T, one gets *a priori* different realizations of the theory of relativity (in its abstract sense), each one embedded in the previous. Of course, the principle of relativity, together with the specification of the total class of phenomena considered, necessarily implies, for consistency, the uniqueness of the transformation equations connecting inertial reference frames.

It is possible to show that, from the above two postulates, there follow — without any additional hypothesis — all the usual "principles" of SR, *i.e.* the "principle of reciprocity", the linearity of transformations between inertial frames, and the invariance of light speed in vacuum.[1]

Concerning this last point, it can be shown in general that postulates 1 and 2 above imply the existence of an invariant, real quantity, having the dimensions of the square of a speed, whose value must be experimentally determined in the framework of the total class C_T of the physical phenomena.[a] Such an invariant speed depends on the interaction (fundamental, or at least phenomenological) ruling the physical phenomenon considered. Therefore *there is, a priori, an invariant speed for every interaction*, namely, a maximal causal speed for every interaction.

All the formal machinery of SR in the Einsteinian sense (including Lorentz transformations and their implications, and the metric structure of space-time) is simply a consequence of the above two postulates and of the choice, for the total class of physical phenomena C_T, of the class of mechanical and electromagnetic phenomena.

1.2. The Principle of Solidarity

The attempt at including the class of nuclear and subnuclear phenomena in the total class of phenomena for which special relativity holds true is therefore expected to imply a generalization of Minkowski metric, analogously to the generalization from the Euclidean to the Minkowski metric in going from mechanics to electrodynamics.

However, in order to avoid misunderstandings, it must be stressed that such an analogy with the extension of the Euclidean metric has to be understood not in the purely geometric meaning, but rather in the sense (already stressed by Penrose[2]) of Euclidean geometry as a physical theory. We shall discuss such a point in more detail in Chap. 3.

Indeed, the generalized metric is endowed with a dynamical character and is not only a consequence, but also an effective description of (the interaction involved in) the class of phenomena considered. It implements

---

[a]The invariant speed is obviously ∞ for Galilei's relativity, and c (light speed in vacuum) for Einstein's relativity.

therefore a feedback between interactions and space-time structure, already realized for gravitation in General Relativity.

This complies with the *"Principle of Solidarity"* stated by B. Finzi in the form already quoted in the Preface, which can be embodied in the following third principle of Relativity:

3. *Principle of Solidarity* (*PS*): Each class of phenomena (namely, each interaction) determines its own space-time.

The mathematical tool to implement the PS is *a deformation* (in the sense specified later on) of the Minkowski metric. This implies, among the others, new, generalized transformation laws, which admit, as a suitable limit, the Lorentz transformations (just like Lorentz transformations represent a covering of the Galilei–Newton transformations).

1.3. Locality and Interaction

1.3.1. *Locality and Nonlocality*

The Principle of Solidarity can in general apply also to interactions nonlocal and not derivable from a potential. Let us therefore clarify what it is meant by such terms in this book.

Essentially two definitions of nonlocality exist in literature. The first amounts to contradict the so-called *Einstein–Bell locality*, which can be stated as follows:

The elements of physical reality of a system cannot be affected instantaneously at a distance (Einstein)

or

The probability of two measurements performed on events separated by a spacelike interval is simply the product of the probabilities of the two measurements separately (Bell).

It is easy to see that such a nonlocality (of quantum nature) is basically connected to the possibility of superluminal signals.

The second definition is related to the functional dependence of the force. A force is local when it depends on a space-time *point* (or, better, on an *infinitesimal* neighborhood of the point); it is nonlocal when it depends on a whole (*finite*) space-time *region*. In the following, we shall just mean this latter definition whenever using the term nonlocal.

1.3.2. *Potential and Nonpotential Interactions*

Let us stress, however, that "local" interaction (in the sense specified above) and "potential" interaction are *not* synonymous, in general. Once one fixes a space-time point, a local interaction is uniquely determined by an infinitesimal neighborhood of the point, whereas a potential interaction is just determined by the value the potential function takes at the point considered. Notice that the derivability from a potential requires the uniqueness of the potential function on the whole space-time region where the force field is defined.

An example of a nonlocal but potential interaction is provided by an interaction described by the potential

$$V(x_i) = \int \prod_{x_j \in \{x\}, j \neq i} dx_j V(\{x\}) \tag{1.1}$$

where $\{x\}$ is the set of metric coordinates, the integration can be definite or indefinite (in the latter case, the potential will depend also on the geometry of the integration regions) and $V(\{x\})$ is regular enough to ensure its integrability (for instance, in the Riemann sense).

On the other side, the electromagnetic (e.m.) interaction associated to a magnetic monopole is an example of a local but nonpotential interaction. In this case, due to the presence of the singular Dirac string, the force field of the monopole is irrotational locally but not globally.[b] This implies that the monopole field is described by *many* (in general different) *local potentials*. By the non-uniqueness of the potential, the e.m. interaction of the magnetic monopole is nonpotential, but it is local indeed.

---

[b]In the language of differential geometry, the field of a Dirac magnetic monopole is associated to a differential form which is closed but not globally exact.

CHAPTER 2

DEFORMED MINKOWSKI SPACE-TIME

As is well known, the Minkowski metric[a]

$$g = \text{diag}(1, -1, -1, -1) \tag{2.1}$$

is a generalization of the Euclidean metric $\epsilon = \text{diag}(1,1,1)$. On the basis of the discussion of Chap. 1, we assume that the metric describes, in an effective way, the interaction, and that there exist interactions more general than the electromagnetic ones (which, as well known, are long-range and derivable from a potential).

The simplest generalization of the space-time metric which accounts for such more general properties of interactions is provided by a *deformation*, η, of the Minkowski metric (2.1), defined as[3-5]

$$\eta = \text{diag}(b_0^2, -b_1^2, -b_2^2, -b_3^2). \tag{2.2}$$

Of course, from a formal point of view metric (2.2) is not new at all. Deformed Minkowski metrics of the same type have already been proposed in the past[6-9] in various physical frameworks, starting from Finsler's generalization of Riemannian geometry[6] to Bogoslowsky's anisotropic space-time[7] to Santilli's isotopic Minkowski space.[8] A phenomenological deformation of the type (2.2) was also obtained by Nielsen and Picek[9] in the context of the electroweak theory. Moreover, although for quite different purposes, "quantum" deformed Minkowski spaces have been also considered in the context of quantum groups.[10] Leaving to later considerations the true specification of the exact meaning of the deformed metric (2.2) in our framework, let us right now stress two basic points.

---

[a]Throughout this book, lower Latin indices take the values $\{1,2,3\}$ and label spatial dimensions, whereas lower Greek indices vary in the range $\{0,1,2,3\}$, with 0 referring to the time dimension. Ordinary 3-vectors are denoted in boldface. Moreover, we adopt the signature $(+, -, -, -)$, and employ the notation "*ESCon*" ("*ESCoff*") to mean that the Einstein sum convention on repeated indices is (is not) used.

1. Firstly, metric (2.2) is supposed to hold at a *local* (and not global) scale, *i.e.* to be valid not everywhere, but only in a suitable (local) space-time region (characteristic of both the system and the interaction considered). We shall therefore refer often to it as a *"topical"* deformed metric.

   In the present case, the term "local" must be understood in the sense that a deformed metric of the kind (2.2) describes the geometry of a 4-dimensional variety attached at a point x of the standard Minkowski space-time, in the same way as a local Lorentz frame is associated (as a tangent space) to each point of the (globally Riemannian) space of Einstein's GR. Another example, on some respects more similar to the present formalism, is provided by a space-time endowed with a vector fibre-bundle structure, where a (maximally symmetric) Riemann space with constant curvature is attached at each point x.[11]

2. Secondly, metric (2.2) is regarded to play a *dynamical role*. So, in order to comply with the solidarity principle, we assume that the parameters b_μ ($\mu = 0, 1, 2, 3$) are, in general, real and positive functions of a given set of observables $\{\mathcal{O}\}$ characterizing the system (in particular, of its total energy, as specified later):

$$\{b_\mu\} = \{b_\mu(\{\mathcal{O}\})\} \in R_0^+, \quad \forall \{\mathcal{O}\}. \tag{2.3}$$

The set $\{\mathcal{O}\}$ represents therefore, in general, a set of non-metric variables ($\{x_{n.m.}\}$).

Equation (2.2) therefore becomes:

$$\begin{aligned}
\eta_{\mu\nu} &= \eta_{\mu\nu}(\{\mathcal{O}\}) \\
&= \text{diag}(b_0^2(\{\mathcal{O}\}), -b_1^2(\{\mathcal{O}\}), -b_2^2(\{\mathcal{O}\}), -b_3^2(\{\mathcal{O}\})) \stackrel{\text{ESC off}}{=} \\
&= \delta_{\mu\nu}(\delta_{\mu 0} b_0^2(\{\mathcal{O}\}) - \delta_{\mu 1} b_1^2(\{\mathcal{O}\}) - \delta_{\mu 2} b_2^2(\{\mathcal{O}\}) - \delta_{\mu 3} b_3^2(\{\mathcal{O}\})). \quad (2.4)
\end{aligned}$$

However, for the moment the deformation of the Minkowski space will be discussed only from a formal point of view, by disregarding the problem of the observables on which the coefficients b_μ actually depend (it will be faced in Chap. 3).

Notice that the first point, *i.e.* the assumed local validity of (2.2), differentiates this approach from those based on Finsler's geometry or from the Bogoslowski's one (which, at least in their standard meaning, do consider deformed metrics at a *global* scale), and makes it similar, on some aspects, to the philosophy and methods of the isotopic generalizations of Minkowski spaces. However, it is well known that Lie-isotopic theories rely in an essential way, from the mathematical standpoint, on (and are strictly characterized by) the very existence of the so-called isotopic unit.[8] In the following, such a formal device will not be exploited (because unessential on all respects), so that the present formalism is not an isotopic one. Moreover, from a physical point of view, the isotopic formalism is expected to apply only to strong interactions.[8] On the contrary, it will be assumed that the (effective) representation of interactions through the deformed metric (2.2) does hold for *all* kinds of interactions (at least for their nonlocal component). In spite of such basic differences this formalism shares some common formal results — as we shall see in the following — with isotopic relativity (like the mathematical expression of the generalized Lorentz transformations).

It is now possible to define a generalized (*"deformed"*) Minkowski space $\tilde{M}(x, \eta(\{\mathcal{O}\}))$ with the same local coordinates x of M (the four-vectors of the usual Minkowski space), but with metric given by the metric tensor η (2.4). The generalized interval in $\tilde{M}$ is therefore given by ($x^\mu = (x^0, x^1, x^2, x^3) = (ct, x, y, z)$, with c being the usual light speed in vacuum) (ESC on)[3–5]:

$$
\begin{aligned}
ds^{\tilde{2}}(\{\mathcal{O}\}) &\equiv b_0^2(\{\mathcal{O}\})c^2 dt^2 - b_1^2(\{\mathcal{O}\})(dx^1)^2 - b_2^2(\{\mathcal{O}\})(dx^2)^2 - b_3^2(\{\mathcal{O}\})(dx^3)^2 \\
&= \eta_{\mu\nu}(\{\mathcal{O}\})dx^\mu dx^\nu = dx * dx.
\end{aligned}
\tag{2.5}
$$

The last step in (2.5) defines the scalar product $*$ in the deformed Minkowski space $\tilde{M}$. Moreover, according to (2.5), we shall use the following notation for the deformed square norm of a four-vector:

$$
|x|_*^2 \equiv x * x = \eta_{\mu\nu}(\{\mathcal{O}\})x^\mu x^\nu = x^{\tilde{2}}.
\tag{2.5'}
$$

In order to evidence some preliminary implications of metric (2.4), let us consider (for simplicity sake and without loss of generality) an isotropic 3-dimensional space, *i.e.*

$$
b_1^2(\{\mathcal{O}\}) = b_2^2(\{\mathcal{O}\}) = b_3^2(\{\mathcal{O}\}) \equiv b^2(\{\mathcal{O}\})
\tag{2.6}
$$

so that the corresponding deformed metric reads

$$
\begin{aligned}
\eta_{\mu\nu_{\text{ISO}}}(\{\mathcal{O}\}) &= \operatorname{diag}(b_0^2(\{\mathcal{O}\}), -b^2(\{\mathcal{O}\}), -b^2(\{\mathcal{O}\}), -b^2(\{\mathcal{O}\})) \stackrel{\text{ESC off}}{=} \\
&= \delta_{\mu\nu}[\delta_{\mu 0}b^2(\{\mathcal{O}\}) - (\delta_{\mu 1} + \delta_{\mu 1} + \delta_{\mu 1})b^2(\{\mathcal{O}\})].
\end{aligned}
\tag{2.7}
$$

One gets, for null separation $ds^{\tilde{2}} = 0$:

$$
\begin{aligned}
ds^{\tilde{2}}(\{\mathcal{O}\}) = 0 &\Leftrightarrow b_0^2(\{\mathcal{O}\})c^2 dt^2 - b^2(\{\mathcal{O}\})[(dx^1)^2 + (dx^2)^2 + (dx^3)^2] = 0 \\
&\Leftrightarrow \frac{(dx^2 + dy^2 + dz^2)}{dt^2} = c^2 \frac{b_0^2(\{\mathcal{O}\})}{b^2(\{\mathcal{O}\})}.
\end{aligned}
\tag{2.8}
$$

From Eq. (2.8) it is easily seen that the *maximal causal speed* in the generalized Minkowski space is given by:

$$u = \frac{b_0(\{\mathcal{O}\})}{b(\{\mathcal{O}\})} c. \tag{2.9}$$

It is worth noticing that a similar result (namely, a "maximum attainable speed", *a priori* different for different physical processes) was also obtained by Coleman and Glashow,[12] in the framework of a discussion of possible effects breaking Lorentz invariance (essentially on a local scale).

Let us remark that u depends explicitly on the metric parameters b_μ, which are *a priori* different for every physical system. However, since the deformation of the metric represents, on average, the effects of the nonlocal interactions involved, it is expected that physical systems with the same kind of interactions (besides the electromagnetic ones) are described by metric parameters of the same order of magnitude (or, at least, this holds true for the ratio b_0/b). In this sense it is possible to refer to u as a "speed of interaction", rather than "speed of the physical system" considered (of course, at the same energy scale).

In Eq. (2.9), the value of u is parametrized in terms of c, and depends on the physical system (and its interactions). Moreover, it is

$$u \gtreqless c \Leftrightarrow \frac{b_0(\{\mathcal{O}\})}{b(\{\mathcal{O}\})} \gtreqless 1. \tag{2.10}$$

In other words, there may be maximal causal speeds either *subluminal* or *superluminal*, depending on the interaction considered.

It is worth to recall that the deformation of the metric, resulting in the interval (2.5), represents a geometrization of a suitable space-time region (corresponding to the physical system considered) that describes, in the average, the effect of nonlocal interactions on a test particle. It is clear that there exist infinitely many deformations of the Minkowski space (precisely, ∞^4), corresponding to the different possible choices of the parameters b_μ, *a priori* different for each physical system.

Moreover, since the usual, "flat" Minkowski metric g (2.1) is related in an essential way to the electromagnetic interaction, we shall always mean in the following — unless otherwise specified — that electromagnetic interactions imply the presence of a fully Minkowskian metric. Actually, as it will be seen, a deformed metric of the type (2.7) is required if one wants to account for possible nonlocal electromagnetic effects (like those occurring in the superluminal wave propagation in waveguides: see Chap. 7).

CHAPTER 3

DESCRIPTION OF INTERACTIONS BY ENERGY-DEPENDENT METRICS

3.1. Energy and the Finzi Principle

Now one has to go into the question of the dependence of the metric parameters $b_\mu(\{\mathcal{O}\})$ on the set of observables $\{\mathcal{O}\}$ of the system considered, and to examine closely the physical meaning of such a functional dependence.

To this aim, the basic question to be put is how to implement Finzi's Principle of Solidarity for all interactions on a mere special-relativistic basis. At present, General Relativity (GR) is the only successful theoretical realization of geometrizing an interaction (the gravitational one). As is well known, energy plays a fundamental role in GR, since the energy-momentum tensor of a given system is the very source of the gravitational field.

A moment's thought shows that this occurs actually also for other interactions. Let us consider, for instance, the case of Euclidean geometry in its intrinsic meaning of a theory of the physical reality at its basic classical (macroscopic) level. Actually, it describes in a quantitative way, in mathematical language, the relations among measured physical entities — distances, in this case —, and therefore of the physical space in which phenomena occur.

However, the measurement of distances depends on the motion of the body which actually performs the measurement. Such a dependence is indeed not on the *kind* of motion, but rather on the *energy* needed to let the body move, and on the *interaction* providing such energy. The measurement of time needs as well a periodic motion with constant frequency, and therefore it too depends on the energy and on the interaction.

This simple example shows how *energy does play a fundamental role in determining the very geometrical structure of space-time.*

Generalizing such an argument, we can state that exchanging energy between particles amounts to measure operationally their space-time separation.[a] Of course such a process depends on the interaction involved in the energy exchange; moreover, each exchange occurs at the maximal causal speed characteristic of the given interaction. It is therefore natural to assume that the measurement of distances, performed by the energy exchange according to a given interaction, realizes the "solidarity principle" between space-time and interactions at the microscopic scale. This allows one to identify the total energy E of the physical process considered as the observable on which the coefficients $b_\mu(\{\mathcal{O}\})$ depend:

$$\{\mathcal{O}\} \equiv E \Leftrightarrow \{b_\mu(\{\mathcal{O}\})\} \equiv \{b_\mu(E)\}, \quad \forall \mu = 0, 1, 2, 3. \qquad (3.1)$$

Actually, since all the functions $\{b_\mu\}$ are dimensionless, they must depend on a dimensionless variable. Then, one has to divide the energy E by a constant E_0 (in general characteristic of each fundamental interaction), with dimensions of energy, so that:

$$\{b_\mu(\{\mathcal{O}\})\} \equiv \left\{b_\mu\left(\frac{E}{E_0}\right)\right\}, \quad \forall \mu = 0, 1, 2, 3. \qquad (3.2)$$

As it will be seen, E_0 has the meaning of a "threshold energy".

Thus, the distance measurement is accomplished by means of the deformed metric tensor, given explicitly by

$$\eta_{\mu\nu}(E) = \mathrm{diag}(b_0^2(E), -b_1^2(E), -b_2^2(E), -b_3^2(E)) \overset{\mathrm{ESC\ off}}{=}$$
$$= \delta_{\mu\nu}(\delta_{\mu 0}b_0^2(E) - \delta_{\mu 1}b_1^2(E) - \delta_{\mu 2}b_2^2(E) - \delta_{\mu 3}b_3^2(E)). \qquad (3.3)$$

Any interaction can be therefore phenomenologically described by metric (3.3) in an *effective* way. This is true in general, but necessary in the case of nonlocal and nonpotential interactions. For force fields which admit a potential, such a description is complementary to the actual one.[b]

One is therefore led to put forward a revision of the concept of "geometrization of an interaction": each interaction produces its own metric, formally expressed by the metric tensor (3.3), but realized via different choices of the set of parameters $b_\mu(E)$. Otherwise said, the $b_\mu(E)$'s are peculiar to every given interaction. The statement that (3.3) provides us with a metric description of an interaction must be just understood in such a sense.

---

[a]Notice that, in this framework, a space-time point has only a mathematical (geometrical) meaning, since it physically corresponds to an energy insufficient to the motion (for the interaction considered). See also Sec. 11.2.
[b]As we shall see in Chap. 8, an example is just provided by the gravitational interaction in the Newtonian limit.

Therefore, the energy-dependent deformation of the Minkowski metric implements a generalization of the concept of geometrization of an interaction (in accordance with Finzi's principle). The GR theory implements a geometrization (at a *global* scale) of the gravitational interaction, based on its derivability from a potential and on the equivalence between the inertial mass of a body and its "gravitational charge". The formalism of energy-dependent metrics allows one instead to implement a geometrization (at a *local* scale) of any kind of interaction characterized by a phenomenology experimentally measurable. As already stressed before, such a formalism applies, in principle, to both fundamental and phenomenological interactions, either potential (gravitational, electromagnetic) or nonpotential (strong, weak), *local* and *nonlocal* (in the sense already specified), for which either an Equivalence Principle holds (as it is the case of gravitation) or (in the more general case) the inertial mass of the body *is not* in general proportional to its charge in the force field considered (e.m., strong, and weak interaction). The basic point of the present way of geometrizing an interaction (thus realizing the Finzi legacy) consists in a "upsetting" of the space-time-energy parametrization. Whereas for potential interactions there exists a potential energy depending on the space-time metric coordinates, one has here to do with a deformed metric tensor η, whose coefficients depend on the energy, that thus assumes a *dynamical* role.

3.2. Energy as Dynamical Variable

From the physical point of view, the energy E is to be understood as the measured energy of the system, and must be therefore regarded as a merely phenomenological variable. As is well known, all the physically realizable detectors work via their electromagnetic interaction in the usual Minkowski space. This is why, in this formalism, the Minkowski space and the e.m. interaction do play a fundamental role. The former is, as already stressed, the cornerstone on which to build up the generalization of Special Relativity based on the deformed metric (3.3). The latter is the comparison term for all fundamental interactions. Let us recall that they are strictly interrelated, since it is just electromagnetism which determines the Minkowski geometry.

From the mathematical standpoint, E is to be considered as a dynamical variable, because it specifies the dynamical behavior of the process under consideration, and, via the metric coefficients, it provides us with a dynamical map — in the energy range of interest — of the interaction ruling the given process.

Let us notice that metric (3.3) plays, for nonpotential interactions, a role analogous to that of the Hamiltonian H for a potential interaction. In particular, the metric tensor η as well is not an input of the theory, but must be built up from the experimental knowledge of the physical data of the system concerned (in analogy with the specification of the Hamiltonian of a potential system). However, there are some differences between η and H worth to be stressed. Indeed, as is well known, H represents the total energy E_{tot} of the system irrespective of the value of E_{tot} and the choice of the variables. On the contrary, $\eta(E)$ describes the variation in the measurements of space and time, in the physical system considered, as E_{tot} changes; therefore, η does depend on the numerical value of H, but not on its functional form. The explicit expression of η depends only on the interaction involved.

It is moreover worth recalling that the use of an energy-dependent space-time metric can be traced back to Einstein himself, who generalized the Minkowski interval as follows

$$ds^2 = \left(1 + \frac{2\phi}{c^2}\right) c^2 dt^2 - (dx^2 + dy^2 + dz^2) \tag{3.4}$$

(where ϕ is the Newtonian gravitational potential), in order to account for the modified rate of a clock in presence of a (weak) gravitational field.

One may be puzzled about the dependence of the metric on the energy, which is not an invariant under usual Lorentz transformations, but transforms like the time-component of a four vector.

Actually, energy is to be regarded, in this formalism, from two different points of view. One has, on one side, the energy as measured in full Minkowskian conditions, which, as such, behaves as a genuine four-vector under usual Lorentz transformations (in the sense that it changes in the usual way if we go, say, from the laboratory frame to another frame in uniform motion with respect to it). Once fixed the frame, one gets a measured value of the energy for a given process (for instance, the energy of the Bose–Einstein correlation phenomenon in pion production, as measured at CERN by the UA1 collaboration). This is the value which enters, as a parameter, in the expression (3.3) of the deformed metric. Such an energy, therefore, is no longer to be considered as a four vector in the deformed Minkowski space, but it is just a quantity whose value determines the deformed geometry of the process considered (or, otherwise speaking, which selects the deformed space-time we have to use to describe the phenomenon).

Let us briefly discuss the phenomenological aspects of the metric dependence on energy. Notice that, in particle collisions, the energy of the incident particle in the laboratory frame, $E = E_L$, can be related to the Mandelstam invariant s (corresponding to the total energy of the colliding particles in the center-of-mass reference frame) by means of the relation, valid at sufficiently high energies:

$$s \approx 2E_L M, \tag{3.5}$$

(where M is the mass of the target),[c] and analogous to the formula relating the laboratory momentum p_L and the invariant flux ϕ: $\phi = p_L M$.

Therefore, in particle collisions, the metric parameters can be indeed considered as dependent on the invariant s:

$$\{b_\mu(E_L)\} \approx \left\{ b_\mu \left(\frac{s}{2M} \right) \right\} \equiv \{\tilde{b}_\mu(s)\}. \tag{3.6}$$

Actually, s is a scalar under usual Lorentz transformations, and, in general, an usual relativistic invariant is no longer unchanged for transformations preserving the deformed interval (2.5). However, let us recall that the energy chosen as phenomenological parameter is that measured electromagnetically, and therefore in presence of a Minkowski metric. This interpretation is supported by the case of colliding beam reactions with different energies: indeed, it would be impossible, otherwise, to define what energy must be used as parameter. Such a point of view can be adopted e.g. in the analysis of the so called "ramping run" of UA1 in order to extract the parameters of the hadronic metric (as functions of the energy) from the experimental data (see Part III).

---

[c]Indeed, the Mandelstam invariant s for two colliding particles is defined as $s \equiv (p_1+p_2)^2$, where p_i is the 4-momentum of the ith particle. In their C.M. frame, $\underline{p}_1 + \underline{p}_2 = 0$, and $s = (E_1 + E_2)^2$. In the laboratory frame, where the target particle 2 of mass M is at rest, the 4-momenta of the two particles are:

$$p_1^\mu = \left(\frac{E_1}{c} = \sqrt{|\mathbf{p}_1|^2 + m^2 c^2}, \mathbf{p}_1 \right), \quad p_2^\mu = (Mc, \mathbf{0}).$$

Therefore

$$s = \frac{E_1^2}{c^2} + M^2 c^2 + 2E_1 M - |\mathbf{p}_1|^2 = |\mathbf{p}_1|^2 + m^2 c^2 + M^2 c^2 + 2E_1 M - |\underline{p}_1|^2$$
$$= m^2 c^2 + M^2 c^2 + 2E_1 M.$$

In the (ultra) relativistic limit $E_1 \gg mc^2$, $E_1 \gg Mc^2$, we get therefore

$$s \approx 2E_1 M$$

which coincides with Eq. (3.5) on account of the fact that $E_1 \equiv E_L$.

In other cases, the phenomenological energy parameter may not be so easy to identify, neither it can be directly related to an invariant quantity. This is the case, for instance, of the lifetime of unstable particles.

Let us explicitly stress that the theory of SR based on metric (2.4) has nothing to do with General Relativity. Indeed, in spite of the formal similarity between the interval (2.5), with the b_μ functions of the coordinates, and the metric structure of a Riemann space, in this framework no mention at all is made of the equivalence principle between mass and inertia, and among non-inertial, accelerated frames. Moreover, General Relativity describes geometrization on a large-scale basis, whereas the special relativity with topical deformed metric describes local (small-scale) deformations of the metric structure (although the term "small scale" must be referred to the real dimensions of the physical system considered).

But the basic difference is provided by the fact that actually the deformed Minkowski space $\tilde{M}$ has zero curvature, as it is easily seen by remembering that, in a Riemann space, the scalar curvature is constructed from the derivatives, with respect to space-time coordinates, of the metric tensor. In others words, the space $\tilde{M}$ is *intrinsically flat* — at least in a mathematical sense. Namely, it would be possible, in principle, to find a change of coordinates, or a rescaling of the lengths, so as to recover the usual Minkowski space. However, such a possibility is only a mathematical, and not a physical one. This is related to the fact that the energy of the process is fixed, and cannot be changed at will. For that value of the energy, the metric coefficients do possess values different from unity, so that the corresponding space $\tilde{M}$, for the given energy value, is actually different from the Minkowski one. The usual space-time M is recovered for a special value E_0 of the energy (characteristic of any interaction), such that indeed

$$\eta(E_0) = g = \text{diag}(1, -1, -1, -1). \tag{3.7}$$

Such a value E_0 (which must be derived from the phenomenology) will be referred to as *the threshold energy of the interaction considered.* As we shall see, it is just the constant introduced in Eq. (3.2) by dimensional arguments.

It is worth stressing that, due to the dependence on energy of the metric parameters b_μ, the maximal causal speed of any interaction, too, is a function of the energy, according to the relation

$$u(E) = \frac{b_0(E)}{b(E)}c. \tag{3.8}$$

However, it remains invariant — for fixed energy values — under generalized Lorentz transformations from a given reference frame to another. Clearly, in Eq. (3.8), the light speed in vacuum, c, does merely play the role of a phenomenological parameter on which the value of u depends.

The maximal causal speed u can be interpreted, from a physical standpoint, as the speed of the quanta of the interaction which requires a representation in terms of a generalized Minkowski space. Since these quanta are associated to lightlike world-lines in $\tilde{M}$ (see Eq. (2.8)), they must be zero-mass particles (with respect to the interaction considered), in analogy with photons (with respect to the e.m. interaction) in the usual SR.[d]

Let us clarify the latter statement. The carriers of a given interaction propagating with the speed u typical of that interaction *are expected to be strictly massless only inside the space whose metric is determined by the interaction considered*. A priori, nothing forbids that such "deformed photons" may acquire a nonvanishing mass in a deformed Minkowski space related to a different interaction.

This might be the case of the massive bosons W^+, W^- and Z^0, carriers of the weak interaction, which would therefore be massless in the space $\tilde{M}(\eta_{\text{weak}}(E))$ related to the weak interaction, but would acquire a mass when considered in the standard Minkowski space M of SR (that, as already stressed, is strictly connected to the electromagnetic interaction,[e] ruling the operation of the measuring devices). In this framework, therefore, it is not necessary to postulate a "symmetry breaking" mechanism (like the Goldstone one in gauge theories) to let particles acquire mass.[f] Mass itself would assume a *relative nature*, related not only to the interaction concerned, but also to the metric background where one measures the energy of the physical system considered. This can be seen if one takes into account the fact in general, for relativistic particles, mass is the invariant norm of 4-momentum, and what is usually measured is not the value of such an invariant, but of the related energy. As it will be seen in Chap. 14, it is possible indeed, in this framework, to give a geometrical meaning to the electron mass, and relate it to the breakdown of local Lorentz invariance.

---

[d] However, the problem of mass requires to be considered in more detail (see Chap. 14).
[e] Actually, as we shall see in Part III, this is strictly true only for $E > E_{0,\text{e.m.}}$, where $E_{0,\text{e.m.}} \simeq 4.5\,\mu\text{eV}$ is the threshold energy for the e.m. interaction, i.e.

$$\tilde{M}(g_{\text{e.m.}}(E))|_{E \gtrsim 4.5\,\mu\text{eV}} = M.$$

[f] On the contrary, if one could build up measuring devices based on interactions different from the e.m. one, the photon might acquire a mass with respect to such a non-e.m. background.

To end this chapter, let us notice that the deformation (3.3) of the Minkowski metric is expected to apply to the description not only of extended particles, but also of quantum pointlike particles, as far as their energy is such that one cannot neglect their associated cloud of virtual quanta.

The problem of a metric description of a given interaction is thus formally reduced to the determination of the coefficients $b_\mu(E)$ from the data on some physical system, whose dynamical behaviour is ruled by the interaction considered.

PART II

RELATIVITY IN A DEFORMED SPACE-TIME

CHAPTER 4

GENERALIZED PRINCIPLE OF RELATIVITY AND LORENTZ TRANSFORMATIONS

4.1. Deformed Special Relativity

In order to develop the relativity theory in a deformed Minkowski space-time, one has to suitably generalize and clarify the basic concepts which are at the very foundation of SR.

Let us first of all define a "topical inertial frame":

(i) a *topical "inertial" frame* (TIF) is a reference frame in which space-time is homogeneous, but space is not necessarily isotropic.

Then, a *"generalized principle of relativity"* (or *"principle of metric invariance"*) can be stated as follows:

(ii) all physical measurements within every topical "inertial" frame must be carried out via the *same* metric.

We shall call *"Deformed Special Relativity"* (DSR) the generalization of SR based on the above two postulates, and whose space-time structure is given by the deformed Minkowski space $\tilde{M}$ introduced in Chap. 2.

4.2. Deriving the Deformed Lorentz Transformations

It follows from the above points (i), (ii) that the transformation equations connecting topical "inertial" frames, *i.e.* the generalized Lorentz transformations, are those which leave invariant the deformed metric when passing from a topical "inertial" frame K, to another frame K', moving with constant velocity with respect to K. Then, physical laws are to be covariant with respect to such generalized transformations.

In other words, the generalized Lorentz transformations are the isometries of the deformed Minkowski space $\tilde{M}$. We shall refer to them in the following as *deformed Lorentz transformations* (DLT). Their explicit form can be derived by the same procedure followed in order to find the Lorentz transformations in the usual Minkowski space.

Consider two TIF, K and K'; by definition, the DLT's leave invariant the deformed interval (2.5), *i.e.*

$$b_0^2 c^2 t^2 - b_1^2 x^2 - b_2^2 y^2 - b_3^2 z^2 = b_0^2 c^2 t'^2 - b_1^2 x'^2 - b_2^2 y'^2 - b_3^2 z'^2. \qquad (4.1)$$

Moreover, it can be assumed, without loss of generality, that the frames K and K' are in standard configuration (*i.e.* their spatial frames coincide at $t = t' = 0$). By choosing the boost direction along $\hat{x}^1 = \hat{x}$, we have therefore $y' = y$, $z' = z$ and Eq. (4.1) reduces to

$$b_0^2 c^2 t^2 - b_1^2 x^2 = b_0^2 c^2 t'^2 - b_1^2 x'^2. \qquad (4.2)$$

From space-time homogeneity it follows that the functional relations between the two sets of coordinates $\{x, y, z, t\}$ and $\{x', y', z', t'\}$ must be linear. Then, in general, the deformed coordinate transformations are to be searched in the form

$$\begin{cases} x' = A_{11}x + A_{14}t \\ y' = y \\ z' = z \\ t' = A_{41}x + A_{44}t \end{cases} \qquad (4.3)$$

where the coefficients $A_{11}, A_{14}, A_{41}, A_{44}$ depend *a priori* in general on $\mathbf{v}$ and $\hat{x}$ (and, parametrically, on the energy).

Notice that the origin O' of TIF K' must move in K with velocity $\mathbf{v} = v^1\hat{x}$, and therefore:

$$x' = 0, \quad x = vt \Leftrightarrow A_{14} = -vA_{11} \Leftrightarrow x' = A_{11}(x - vt). \qquad (4.4)$$

Replacing (4.3), (4.4) in (4.2) yields

$$b_0^2 c^2 t^2 - b_1^2 x^2 = b_0^2 c^2 (A_{41}x + A_{44}t)^2 - A_{11}^2 b_1^2 x^2 (x - vt)^2 \qquad (4.5)$$

which implies the following 3×3 quadratic system:

$$\begin{cases} c^2 = c^2 A_{44}^2 - \left(\dfrac{b_1}{b_0}\right)^2 A_{11}^2 v^2 \\[2mm] -1 = c^2 \left(\dfrac{b_0}{b_1}\right)^2 A_{41}^2 - A_{11}^2 \\[2mm] 0 = c^2 \left(\dfrac{b_0}{b_1}\right)^2 A_{41} A_{44} + A_{11}^2 v \end{cases} \qquad (4.6)$$

with general solution

$$A_{11} = A_{44} = \pm \left(1 - \left(\frac{vb_1}{cb_0}\right)^2\right)^{-1/2} \tag{4.7}$$

$$A_{41} = \mp \left(\frac{vb_1^2}{c^2b_0^2}\right) \left(1 - \left(\frac{vb_1}{cb_0}\right)^2\right)^{-1/2} = -\left(\frac{vb_1^2}{c^2b_0^2}\right) A_{11}. \tag{4.8}$$

The final result is

$$\begin{cases} x' = \tilde{\gamma}(x - vt) = \tilde{\gamma}\left(x - \tilde{\beta}\frac{b_0}{b_1}ct\right) \\ y' = y \\ z' = z \\ t' = \tilde{\gamma}\left(t - \frac{vb_1^2}{c^2b_0^2}x\right) = \tilde{\gamma}\left(t - \frac{\tilde{\beta}^2}{v}x\right) \end{cases} \tag{4.9}$$

where v is the relative speed of the reference frames, and

$$\tilde{\beta} = \frac{v}{u}; \tag{4.10}$$

$$\tilde{\gamma} = (1 - \tilde{\beta}^2)^{-1/2}. \tag{4.11}$$

Transformations (4.9) do formally coincide with the isotopic Lorentz transformations. However, in the present context their physical meaning is different, as it is easily seen e.g. by the identification of the maximal causal speed u with the speed characteristic of the quanta of a given interaction (see Sec. 3). In particular, the parametrization (4.10) of the deformed velocity parameter $\tilde{\beta}$ in terms of u immediately shows that is always $\tilde{\beta} < 1$, so that $\tilde{\gamma}$ never takes imaginary values (contrarily to the isotopic case). Moreover, no reference at all is made, in this framework, to the existence of an underlying "medium".

It must be carefully noted that, like the metric, also the generalized Lorentz transformations depend on the energy. This means that one gets different transformation laws for different values of E, but still with the same functional dependence on the energy, so that the invariance of the deformed interval (2.5) is always ensured (provided that the process considered does always occur via the same interaction).

Indeed, the energy E can be considered fixed also because, from a quantum point of view, energy can be transferred only by finite amounts.

Differentiating Eqs. (4.9), we get therefore

$$\begin{cases} udt' + t'du = \tilde{\gamma}(udt - \tilde{\beta}dx) + [d\tilde{\gamma}(ut - \tilde{\beta}x) + \tilde{\gamma}(tdu - xd\tilde{\beta}]; \\ dx' = \tilde{\gamma}(dx - \tilde{\beta}udt) + [d\tilde{\gamma}(x - \tilde{\beta}ut) - \tilde{\gamma}(t\tilde{\beta}du + td\tilde{\beta})], \end{cases} \tag{4.12}$$

where, by the above argument, $dE = 0$ and therefore $d\tilde{\gamma} = d\tilde{\beta} = du = 0$. Squaring (4.12) and subtracting, we find

$$dx'^2 - u^2 dt'^2 = \tilde{\gamma}^2[(dx - \tilde{\beta}dt)^2 - (udt - \tilde{\beta}dx)^2] = dx^2 - u^2 dt^2 \tag{4.13}$$

where in the last step use has been made of Eq. (4.2). Exploiting the explicit expression of u, Eq. (2.9), one has finally

$$ds'^2 = ds^2, \tag{4.14}$$

i.e. the deformed Lorentz transformations (4.9) are actually the isometries of the deformed Minkowski space $\tilde{M}$, in spite of their dependence on the energy.

As already stressed in Chap. 2 (see Eq. (2.10)), it is possible to have $u \geq c$ and therefore $c \leq v \leq u$, *i.e.* superluminal motions are allowed. Let us remark that the possibility of tachyonic speeds is accomplished, within this framework, without any recourse neither to imaginary quantities nor to singularities in the transformation laws (unlike the standard case), because it is always $v \leq u$ (even if $v \geq c$), so that the relativistic factor $\tilde{\gamma}$ (Eq. (4.11)) takes only real values, as already noted above.

4.3. Maximal Causal Speed Revisited

In $\tilde{M}$, it is possible *a priori* to consider two scalar products between 3-vectors $\mathbf{v}_1, \mathbf{v}_2$: the standard, Euclidean one $\cdot$, defined by means of the metric tensor $g_{ik} = \delta_{ik}$, and the deformed one, induced by the deformed scalar product $*$ in $\tilde{M}$, and defined by means of the metric tensor $-\eta_{ik}(E) \overset{\text{ESCoff}}{=} b_i^2(E)\delta_{ik}$ (where the sign $-$ is obviously introduced in order to get a positive 3-vector norm) as follows (cf. Eq. (2.5)):

$$\mathbf{v}_1 * \mathbf{v}_2 \equiv -\sum_{i=1}^{3} \eta_{ij}(E)(v_1)^i(v_2)^j$$

$$= \sum_{i=1}^{3} b_i^2(E)\delta_{ij}(v_1)^i(v_2)^j$$

$$= b_1^2(E)(v_1)^1(v_2)^1 + b_2^2(E)(v_1)^2(v_2)^2 + b_3^2(E)(v_1)^3(v_2)^3. \tag{4.15}$$

In the following, $|\mathbf{v}|_*$ will denote the absolute value of a 3-vector with respect to the deformed scalar product $*$ (cf. Eq. (2.5′)), whereas the notation $|\mathbf{v}| = v$ will be used for the norm of $\mathbf{v}$ with respect to the standard product $\cdot$.

As is well known, the maximal causal speed in M is obtained by putting $ds^2 = 0$, whence

$$ds^2 = 0 \Leftrightarrow c^2 dt^2 - dx^2 - dy^2 - dz^2 = 0 \Leftrightarrow \frac{dx^2 + dy^2 + dz^2}{dt^2} = c^2. \qquad (4.16)$$

Then one interprets c as the maximal causal speed along any direction of the (Euclidean) space R^3 (embedded in the pseudo-euclidean Minkowski space-time M). Such an interpretation is obviously based on the physical fact that c coincides with the light speed in vacuum, and on the isotropy of R^3. Therefore c represents the value of any of the three components of the maximal causal velocity vector (m.c.v.) of SR, $\mathbf{u}_{SR}$, namely:

$$\mathbf{u}_{SR} = (c, c, c). \qquad (4.17)$$

Then, c^2 is not, in general, a square modulus, but the square of any component of $\mathbf{u}_{SR}$, whose square modulus (with respect to the Euclidean scalar product $\cdot$), is instead:

$$|\mathbf{u}_{SR}|^2 \equiv \sum_{i=1}^{3} (u_{SR}^i)^2 = 3c^2 \qquad (4.18)$$

so that

$$u_{SR}^i = \frac{1}{\sqrt{3}} |\mathbf{u}_{SR}| \quad \forall i = 1, 2, 3. \qquad (4.19)$$

The above procedure must be suitably modified in the DSR case, due to the space anisotropy of $\tilde{M}$.

Actually, in order to sort out a single component of the 3-vector m.c.v., in a general 4-d special-relativistic theory (characterized by a diagonal metric tensor $g_{\mu\nu}(\{\mathcal{O}\})$, where $\{\mathcal{O}\}$ is a set of observables corresponding to non-metric variables), one has to exploit a "*directional separation*" (or "*dimensional separation*") method, which consists of the following three-step recipe (ESCoff throughout):

1. Set $d\tilde{s}^2$ equal to zero:

$$d\tilde{s}^2 = 0 \Leftrightarrow g_{00}(\{\mathcal{O}\})c^2 dt^2 + \sum_{i=1}^{3} g_{ii}(\{\mathcal{O}\})(dx^i)^2 = 0. \qquad (4.20)$$

2. In order to find the *i*th component $u^i(\{\mathcal{O}\})$ of the m.c.v., put $dx^j = 0$ $(j \neq i)$, thus getting

$$g_{00}(\{\mathcal{O}\})c^2 dt^2 + g_{ii}(\{\mathcal{O}\})(dx^i)^2 = 0. \tag{4.21}$$

3. Evidence on the l.h.s. of (4.21) a quantity with physical dimensions $\frac{[\text{space}]}{[\text{time}]} = [\text{velocity}]$; at this point, we have two different subcases:

(I) One carries to the l.h.s. of (4.21) $\frac{dx^i}{dt}$ (which amounts to consider the 3-d Euclidean product ·), thus getting an *anisotropic* m.c.v.:

$$u^i(\{\mathcal{O}\}) \equiv \frac{dx^i}{dt} = \frac{(g_{00}(\{\mathcal{O}\}))^{1/2}}{(-g_{ii}(\{\mathcal{O}\}))^{1/2}}c \quad \forall\, i = 1, 2, 3. \tag{4.22}$$

(II) One carries to the l.h.s. of (4.21) $(-g_{ii}(\{\mathcal{O}\}))^{1/2}\frac{dx^i}{dt}$ (which amounts to consider the 3-d deformed product $*$ defined by $-g_{ij}(\{\mathcal{O}\}) = \delta_{ij}|g_{ii}(\{\mathcal{O}\})|$, thus getting an *isotropic* m.c.v.:

$$u^i(\{\mathcal{O}\}) \equiv (-g_{ii}(\{\mathcal{O}\}))^{1/2}\frac{dx^i}{dt} = (g_{00}(\{\mathcal{O}\}))^{1/2}c \quad \forall\, i = 1, 2, 3. \tag{4.23}$$

The two subcases I and II differ essentially by the different way of implementing the space anisotropy. In the former case, the anisotropy is embedded in the definition of m.c.v.; in the latter one, in the scalar product.[a]

Specializing the above equations to the DSR framework, we get therefore, in the two subcases:

(I)
$$u^i_{DSR,I}(E) \equiv u^i(E) = c\frac{b_0(E)}{b_i(E)} \tag{4.24}$$

---

[a]Let us notice that the directionally separating procedure can be consistently applied only to (special- or general relativistic) metrics which are fully diagonal. This is obviously due to the mixings between different space directions which arise in the case of non-diagonal metrics.

Of course, such a procedure gives (in either subcase) the same standard result when applied to SR. In fact:

$$u^i_{SR} = (-g_{ii})^{1/2}\frac{dx^i}{dt} = (g_{00})^{1/2}c = \frac{dx^i}{dt}$$
$$= \frac{(g_{00})^{1/2}}{(-g_{ii})^{1/2}}c = c \quad \forall\, i = 1, 2, 3.$$

$$|\mathbf{u}_{DSR,I}(E)| = \left(\sum_{i=1}^{3}(u^i_{DSR,I}(E))^2\right)^{1/2}$$

$$= cb_0(E)\left(\frac{1}{b_1^2(E)} + \frac{1}{b_2^2(E)} + \frac{1}{b_3^2(E)}\right)^{1/2} \tag{4.25}$$

The vector $\mathbf{u}$ is the (spatially) anisotropic generalization of the maximal causal speed derived in the (spatially) isotropic case, Eq. (3.8);

(II)

$$u^i_{DSR,II}(E) \equiv w^i(E) = cb_0(E) \tag{4.26}$$

$$|\mathbf{u}_{DSR,II}(E)|_* = \left(\sum_{i=1}^{3} b_i^2(E)(u^i_{DSR,II}(E))^2\right)^{1/2}$$

$$= cb_0(E)(b_1^2(E) + b_2^2(E) + b_3^2(E))^{1/2} \tag{4.27}$$

whence

$$u^i_{DSR,II}(E) = (b_1^2(E) + b_2^2(E) + b_3^2(E))^{-1/2}|\mathbf{u}_{DSR,II}(E)|_* \tag{4.28}$$

i.e. in this subcase (unlike the previous one, see Eqs. (4.24) and (4.25)) one can state a proportionality relation by an overall factor (even if dependent on the metric coefficients) between $u^i_{DSR,II}(E)$ and $|\mathbf{u}_{DSR,II}(E)|_*$.

We have therefore shown that the two different procedures of directional separation lead to two different mathematical definitions of maximal causal velocity, an isotropic ($\mathbf{w}$, Eq. (4.26)) and an anisotropic ($\mathbf{u}$, Eq. (4.24)) one. The choice between them must be done on a physical basis (see Subsec. 4.4.3).

4.4. Boosts in DSR

4.4.1. *Boost in a Generic Direction*

In this case, the relative velocity is $\mathbf{v} = v^1\hat{x} + v^2\hat{y} + v^3\hat{z}$, and we have to suitably generalize definitions (4.10), (4.11) as follows:

$$\tilde{\boldsymbol{\beta}} \equiv \frac{\mathbf{v}}{\mathbf{u}} \equiv \left(\frac{v^1 b_1(E)}{cb_0(E)}, \frac{v^2 b_2(E)}{cb_0(E)}, \frac{v^3 b_3(E)}{cb_0(E)}\right) \tag{4.29}$$

$$\tilde{\gamma} \equiv \left(1 - |\tilde{\boldsymbol{\beta}}|^2\right)^{-1/2} \tag{4.30}$$

where (cf. Eq. (4.17))

$$\mathbf{u} = \left(c\frac{b_0(E)}{b_1(E)}, c\frac{b_0(E)}{b_2(E)}, c\frac{b_0(E)}{b_3(E)}\right). \tag{4.31}$$

Notice that $\tilde{\beta} \equiv \frac{\mathbf{v}}{\mathbf{u}} \neq \frac{\mathbf{v}}{u}$. This follows from the anisotropy of the 3-vector $\mathbf{u}$, and it is to be compared with the SR case, where $\beta \equiv \frac{\mathbf{v}}{\mathbf{u}} = \frac{\mathbf{v}}{c}$. In general, it is possible to state that

$$\frac{\mathbf{m}}{\mathbf{n}} = \frac{1}{n}\mathbf{m} \Leftrightarrow \mathbf{n} = (n, n, n)$$

i.e. iff $\mathbf{n}$ is a spatially isotropic 3-vector.

In order to derive the expression of the deformed boost in a generic direction, it is possible to use the same method of the previous case (see Appendix). However, it is simpler to consider the notion of parallelism between 3-vectors in $\tilde{M}(E)^{\mathrm{b}}$ and decompose the space vector $\mathbf{x}$ in two components, $\mathbf{x}_{\parallel}$ and $\mathbf{x}_{\perp}$, parallel and orthogonal, respectively, to the boost direction $\hat{v}$

$$\mathbf{x} = \mathbf{x}_{\parallel} + \mathbf{x}_{\perp} \tag{4.32}$$

$$\mathbf{x}_{\parallel} \equiv \hat{v}(\hat{v} * \mathbf{x}) = \frac{\mathbf{v}}{|\mathbf{v}|_*^2}(\mathbf{v} * \mathbf{x}) = \frac{\mathbf{v}}{\mathbf{v} * \mathbf{v}}(\mathbf{v} * \mathbf{x})$$

$$= \frac{\sum_{i=1}^{3} b_i^2(E)v^i x^i}{\sum_{i=1}^{3} b_i^2(E)(v^i)^2}\mathbf{v} \neq \hat{\tilde{\beta}}(\hat{\tilde{\beta}} * \mathbf{x})$$

$$= \frac{\tilde{\beta}}{|\tilde{\beta}|_*^2}(\tilde{\beta} * \mathbf{x}) = \frac{\tilde{\beta}}{\tilde{\beta} * \tilde{\beta}}(\tilde{\beta} * \mathbf{x}) = \frac{\sum_{i=1}^{3} b_i^2(E)\tilde{\beta}^i x^i}{\sum_{i=1}^{3} b_i^2(E)(\tilde{\beta}^i)^2}\tilde{\beta} \tag{4.33}$$

(with $\|\|_*$ denoting the absolute value of a 3-vector with respect to the deformed scalar product $*$, whereas the notation $\|\|$ will be used for the 3-vector norm with respect to the standard product $\cdot$)

$$x_{\parallel}^i \equiv \frac{\sum_{k=1}^{3} b_k^2(E)v^k x^k}{\sum_{k=1}^{3} b_k^2(E)(v^k)^2}v^i \neq \frac{\sum_{k=1}^{3} b_k^2(E)\tilde{\beta}^k x^k}{\sum_{k=1}^{3} b_k^2(E)(\tilde{\beta}^k)^2}\tilde{\beta}^i \tag{4.34}$$

$$\mathbf{x}_{\perp} \equiv \mathbf{x} - \mathbf{x}_{\parallel} = \mathbf{x} - \frac{\sum_{i=1}^{3} b_i^2(E)v^i x^i}{\sum_{i=1}^{3} b_i^2(E)(v^i)^2}\mathbf{v}$$

$$\neq \mathbf{x} - \frac{\sum_{i=1}^{3} b_i^2(E)\tilde{\beta}^i x^i}{\sum_{i=1}^{3} b_i^2(E)(\tilde{\beta}^i)^2}\tilde{\beta} \tag{4.35}$$

---

[b]The definitions of parallelism and orthogonality are to be meant in the sense of the deformed 3D scalar product $*$ (see Eq. (2.5)).

$$x^i_\perp \equiv x^i - \frac{\sum_{k=1}^3 b_k^2(E)v^k x^k}{\sum_{k=1}^3 b_k^2(E)(v^k)^2} v^i$$

$$\neq x^i - \frac{\sum_{k=1}^3 b_k^2(E)\tilde{\beta}^k x^k}{\sum_{k=1}^3 b_k^2(E)(\tilde{\beta}^k)^2} \tilde{\beta}^i. \tag{4.36}$$

It is easily checked that indeed

$$\mathbf{x} * \mathbf{v} = \sum_{i=1}^3 b_i^2(E)x^i v^i = \frac{\sum_{i=1}^3 b_i^2(E)x^i v^i}{\sum_{i=1}^3 b_i^2(E)(v^i)^2} \sum_{k=1}^3 b_k^2(E)(v^k)^2$$

$$= \frac{\sum_{i=1}^3 b_i^2(E)x^i v^i}{\sum_{i=1}^3 b_i^2(E)(v^i)^2} \mathbf{v} * \mathbf{v} = \mathbf{x}_\parallel * \mathbf{v} = |\mathbf{x}_\parallel|_* |\mathbf{v}|_* \tag{4.37}$$

$$\mathbf{x}_\perp * \mathbf{v} = \mathbf{x} * \mathbf{v} - \mathbf{x}_\parallel * \mathbf{v} = 0. \tag{4.38}$$

Then, applying the boost (4.9) to $\mathbf{x}_\parallel$ and $\mathbf{x}_\perp$ yields

$$\begin{cases} \mathbf{x}'_\parallel = \tilde{\gamma}(\mathbf{x}_\parallel - \mathbf{v}t) \\ \mathbf{x}'_\perp = \mathbf{x}_\perp \\ t' = \tilde{\gamma}\left(t - \sum_{i=1}^3 \frac{v^i b_i^2(E)}{c^2 b_0^2(E)} x^i\right) = \tilde{\gamma}(t - \tilde{\mathbf{B}} \cdot \mathbf{x}) = \tilde{\gamma}(t - \tilde{\mathbf{B}}^{(*)} * \mathbf{x}) \end{cases} \tag{4.39}$$

where we put

$$\tilde{\gamma} \equiv (1 - \tilde{\boldsymbol{\beta}} \cdot \tilde{\boldsymbol{\beta}})^{-1/2} = \left(1 - \tilde{\boldsymbol{\beta}}^{(*)} * \tilde{\boldsymbol{\beta}}^{(*)}\right)^{-1/2}$$

$$= \left[1 - \left(\frac{v^1 b_1(E)}{cb_0(E)}\right)^2 - \left(\frac{v^2 b_2(E)}{cb_0(E)}\right)^2 - \left(\frac{v^3 b_3(E)}{cb_0(E)}\right)^2\right]^{-1/2} \tag{4.40}$$

$$\tilde{\boldsymbol{\beta}}^{(*)} \equiv \frac{\mathbf{v}}{\mathbf{w}} = \left(\frac{v^1}{cb_0(E)}, \frac{v^2}{cb_0(E)}, \frac{v^3}{cb_0(E)}\right) = \frac{1}{cb_0(E)}\mathbf{v} \tag{4.41}$$

$$\mathbf{w} \equiv (cb_0(E), cb_0(E), cb_0(E)) \tag{4.42}$$

$$\tilde{\mathbf{B}} \equiv \frac{\mathbf{v}}{\mathbf{u}^2} = \left(\frac{v^1 b_1^2(E)}{c^2 b_0^2(E)}, \frac{v^2 b_2^2(E)}{c^2 b_0^2(E)}, \frac{v^3 b_3^2(E)}{c^2 b_0^2(E)}\right) \tag{4.43}$$

$$\tilde{\mathbf{B}}^{(*)} \equiv \frac{\mathbf{v}}{\mathbf{w}^2} = \frac{1}{c^2 b_0^2(E)}\mathbf{v}. \tag{4.44}$$

It follows therefore that the deformed boosts admit a double treatment, either:

(I) In terms of the Euclidean scalar product $\cdot$, of the (anisotropic) m.c.v. $\mathbf{u}$ and of the related velocity parameters $\tilde{\boldsymbol{\beta}}$ and $\tilde{\mathbf{B}}$, or

(II) in terms of the deformed product $*$, of the (isotropic) m.c.v. $\mathbf{w}$ and of the related velocity parameters $\tilde{\beta}^{(*)}$ and $\hat{\mathbf{B}}^{(*)}$.[c]

Then, the space vector transforms as:

$$\mathbf{x}' = \mathbf{x}'_{\parallel} + \mathbf{x}'_{\perp} = \tilde{\gamma}(\mathbf{x}_{\parallel} - \mathbf{v}t) + \mathbf{x}_{\perp}$$

$$= \mathbf{x} + (\tilde{\gamma} - 1)\hat{v}(\hat{v} * \mathbf{x}) - \tilde{\gamma}\mathbf{v}t = \mathbf{x} + (\tilde{\gamma} - 1)\frac{\mathbf{v}}{|\mathbf{v}|^2_*}(\mathbf{v} * \mathbf{x}) - \tilde{\gamma}\mathbf{v}t \quad (4.45)$$

and we eventually find the expression of the deformed boost in a generic direction:

$$\begin{cases} \mathbf{x}' = \mathbf{x} + (\tilde{\gamma} - 1)\dfrac{\mathbf{v}}{|\mathbf{v}|^2_*}(\mathbf{v} * \mathbf{x}) - \tilde{\gamma}\mathbf{v}t \\[2mm] t' = \tilde{\gamma}(t - \widetilde{\mathbf{B}} \cdot \mathbf{x}) = \tilde{\gamma}(t - \widetilde{\mathbf{B}}^{(*)} * \mathbf{x}). \end{cases} \quad (4.46)$$

4.4.2. *Symmetrization of Deformed Boosts*

As in the case of standard SR, it is possible to symmetrize the expression of boosts in DSR by introducing suitable time coordinates.

Let us first consider a deformed boost along $\hat{x}^i$ ($i = 1, 2, 3$); the symmetrization transformation (a "dimensionally homogenizing dilato-contraction") of t is given by

$$x^0 \equiv u^i t = c\frac{b_0(E)}{b_i(E)}t; \quad x^{i\prime} \equiv x^i. \quad (4.47)$$

The deformed metric tensor in the new "primed" coordinates, $\{x^{\mu\prime}\} = \{x^0, x, y, z\}$, reads:

$$\eta'_{\mu\nu}(E) \overset{\text{ESC on}}{=} \eta_{\alpha\beta}(E)\frac{\partial x^\alpha}{\partial x^{\mu\prime}}\frac{\partial x^\beta}{\partial x^{\nu\prime}} = \text{diag}(b_i^2(E), -b_1^2(E), -b_2^2(E), -b_3^2(E))$$

$$\overset{\text{ESC off}}{=} \delta_{\mu\nu}[b_i^2(E)\delta_{\mu 0} - b_1^2(E)\delta_{\mu 1} - b_2^2(E)\delta_{\mu 2} - b_3^2(E)\delta_{\mu 3}]. \quad (4.48)$$

Equation (4.9) takes therefore the symmetric form in x^i e x^0 (ESC off):

$$\begin{cases} x^{i\prime} = \tilde{\gamma}(x^i - \tilde{\beta}^i x^0); \\[1mm] x^{k\neq i\prime} = x^{k\neq i}; \\[1mm] x^{0\prime} = \tilde{\gamma}(x^0 - \tilde{\beta}^i x^i). \end{cases} \quad (4.49)$$

Transformation (4.49) does not symmetrize the deformed boost in a generic direction (unlike the case of SR, where the same transformation

---

[c]It is possible to show that, in this case, more equivalent forms of the deformed boost (4.39) exist. As is easily seen, this is due to the fact that, in general, $\hat{\beta} \neq \hat{v}$ and $\hat{B} \neq \hat{v}$, whereas $\widetilde{\beta^{(*)}} = \hat{v} = \widetilde{B^{(*)}}$.

$x^0 = ct$ symmetrizes both boosts). In this case, the symmetrization is possible only if the treatment II (based on the deformed scalar product $*$) is used.

In fact, by using the proportionality (see Eqs. (4.41) and (4.44) and the footnote at p. 36) among $\tilde{\boldsymbol{\beta}}^{(*)}$, $\tilde{\mathbf{B}}^{(*)}$ and $\mathbf{v}$, the following transformation on t (see Eq. (4.42))

$$x^0 \equiv cb_0(E)t = w^k t \ (\forall k = 1, 2, 3); \quad x^{i'} \equiv x^i \ (\forall i = 1, 2, 3) \qquad (4.50)$$

does symmetrize Eq. (4.39) in $\mathbf{x}_\parallel$ e x^0:

$$
\begin{cases}
\mathbf{x}'_\parallel = (1 - \tilde{\boldsymbol{\beta}}^{(*)} * \tilde{\boldsymbol{\beta}}^{(*)})^{-1/2}(\mathbf{x}_\parallel - \tilde{\boldsymbol{\beta}}^{(*)} x^0) \\
\mathbf{x}'_\perp = \mathbf{x}_\perp \\
x^{0'} = \begin{cases}
(1 - \tilde{\boldsymbol{\beta}}^{(*)} * \tilde{\boldsymbol{\beta}}^{(*)})^{-1/2}(x^0 - \tilde{\boldsymbol{\beta}}^{(*)} * \mathbf{x}) \\
= (1 - \tilde{\boldsymbol{\beta}}^{(*)} * \tilde{\boldsymbol{\beta}}^{(*)})^{-1/2}(x^0 - \tilde{\boldsymbol{\beta}}^{(*)} * \mathbf{x}_\parallel).
\end{cases}
\end{cases}
\qquad (4.51)
$$

Under transformation (4.51), the metric tensor becomes:

$$\eta'_{\mu\nu}(E) \overset{\text{ESC on}}{=} \eta_{\alpha\beta}(E)\frac{\partial x^\alpha}{\partial x^{\mu'}}\frac{\partial x^\beta}{\partial x^{\nu'}} = \text{diag}(1, -b_1^2(E), -b_2^2(E), -b_3^2(E))$$

$$\overset{\text{ESC off}}{=} \delta_{\mu\nu}[\delta_{\mu 0} - b_1^2(E)\delta_{\mu 1} - b_2^2(E)\delta_{\mu 2} - b_3^2(E)\delta_{\mu 3}]. \qquad (4.52)$$

Therefore the symmetrization of the deformed boost in a generic direction makes the 4-d metric isochronous, since $\eta'_{00} = 1$ so that $\tau = t$ (namely proper time coincides with coordinate time).

Let us finally notice that, like in the SR case, the boost in generic direction expressed in terms of $\mathbf{x}$ e t (Eq. (4.39)) cannot in general be symmetrized.

Apparently Eqs. (4.9) are asymmetrical in the behaviour of x' and t', unlike the usual Lorentz transformations, which are fully symmetric when putting $x^0 = ct$. However, such asymmetry is only formal. It can be removed by introducing, in analogy with the electromagnetic case, a time coordinate defined in terms of the maximal causal speed u in the generalized Minkowski space considered:

$$x^0 = ut = \left(\frac{b_0}{b}c\right)t \qquad (4.53)$$

and changing the metric tensor η into

$$\eta' = \text{diag}(b^2, -b^2, -b^2, -b^2) = b^2 g. \qquad (4.54)$$

Then, the generalized Lorentz transformations in $\tilde{M}'$ take the symmetrical form

$$\begin{cases} x^{0\prime} = \tilde{\gamma}(x^0 - \tilde{\beta}x^1); \\ x^{1\prime} = \tilde{\gamma}(x^1 - \tilde{\beta}x^0); \\ x^{2\prime} = x^2; \\ x^{3\prime} = x^3. \end{cases} \qquad (4.55)$$

It is easily seen that the deformed Minkowski spaces $\tilde{M}$ and $\tilde{M}'$, with metrics (2.2) and (4.54) respectively, are isometric, because they have the same interval (2.5). They are therefore fully equivalent in every respect, and it is therefore possible to use indifferently either transformation (4.9) or (4.55). The main advantage of the latter ones is that, due to relation (4.53), the formulae holding for $\tilde{M}'$ are immediately got from those of the standard special relativity by simply replacing everywhere c by u.

4.4.3. *Velocity Composition Law in $\tilde{M}$ and the Invariant Maximal Speed*

We have seen in Sec. 4.3 that the directionally separating approach (mandatory in the deformed case) yields two different *mathematical* definitions $\mathbf{u}$ (Eq. (4.24)) and $\mathbf{w}$ (Eq. (4.26)) of maximal causal velocity in DSR. The choice between them must be done on a physical basis, by checking their actual invariance under deformed boosts.

To this aim, one has to derive the generalized velocity composition law valid in $\tilde{M}$. For a deformed boost in the direction $\hat{x}^i$, differentiating the inverse of Eq. (4.9) yields (on account of the fact that $dE = 0$ in DSR) (ESC off):

$$\begin{cases} dx^i = \tilde{\gamma}(dx^{i\prime} + v^i dt') \\ dx^{k \neq i} = dx^{k \neq i\prime} \\ dt = \tilde{\gamma}\left(dt' + \dfrac{v^i b_i^2(E)}{c^2 b_0^2(E)} dx^{i\prime} \right) \end{cases} \qquad (4.56)$$

with $\tilde{\gamma}$ given by (4.11). Since

$$\frac{dx^i}{dt} = v^i, \quad \frac{dx^{i\prime}}{dt'} = v^{i\prime}, \quad \frac{dx^{k \neq i}}{dt} = v^{k \neq i}, \quad \frac{dx^{k \neq i\prime}}{dt'} = v^{k \neq i\prime} \qquad (4.57)$$

one gets the *deformed velocity composition law* (in compact notation, ESC off)

$$v^k = \frac{v^{k\prime} + \delta_{ik}v^i}{\left[1 + \left(\frac{b_i(E)}{b_0(E)}\right)^2 \frac{v^i v^{i\prime}}{c^2}\right]\{\tilde{\gamma}(E) + \delta_{ik}[1 - \tilde{\gamma}(E)]\}}. \qquad (4.58)$$

This relation can be expressed in terms of the standard 3-d scalar product · (and therefore of the anisotropic maximal velocity **u**) (approach I) as

$$\begin{aligned} v^k &= \frac{v^{k\prime} + \delta_{ik}v^i}{\left[1 + \frac{\mathbf{v} \cdot \mathbf{v}'}{(u^i(E))^2}\right]\{\tilde{\gamma}(E) + \delta_{ik}[1 - \tilde{\gamma}(E)]\}} \\ &= \frac{v^{k\prime} + \delta_{ik}v^i}{\left[1 + \frac{\tilde{\boldsymbol{\beta}} \cdot \mathbf{v}'}{u^i(E)}\right]\{\tilde{\gamma}(E) + \delta_{ik}[1 - \tilde{\gamma}(E)]\}} \end{aligned} \qquad (4.59)$$

where

$$\tilde{\beta}^i(E) = \frac{v^i}{u^i(E)}; \quad \tilde{\gamma}(E) = (1 - \tilde{\boldsymbol{\beta}}(E) \cdot \tilde{\boldsymbol{\beta}}(E))^{-1/2}. \qquad (4.60)$$

Alternatively, we can use approach II, based on the deformed scalar product $*$ (and therefore the isotropic maximal velocity **w**) and write Eq. (4.59) as

$$\begin{aligned} v^k &= \frac{v^{k\prime} + \delta_{ik}v^i}{\left[1 + \frac{\mathbf{v} * \mathbf{v}'}{(w^i(E))^2}\right]\{\tilde{\gamma}(E) + \delta_{ik}[1 - \tilde{\gamma}(E)]\}} \\ &= \frac{v^{k\prime} + \delta_{ik}v^i}{\left[1 + \frac{\tilde{\boldsymbol{\beta}}^{(*)} * \mathbf{v}'}{w^i(E)}\right]\{\tilde{\gamma}(E) + \delta_{ik}[1 - \tilde{\gamma}(E)]\}} \end{aligned} \qquad (4.61)$$

with

$$\tilde{\beta}^{(*)i}(E) = \frac{v^i}{w^i(E)}; \quad \tilde{\gamma}(E) = (1 - \tilde{\boldsymbol{\beta}}^{(*)}(E) * \tilde{\boldsymbol{\beta}}^{(*)}(E))^{-1/2}. \qquad (4.62)$$

It is now an easy task to check the truly maximal character of the two velocities. Indeed, if $v^{i\prime} = u^i(E)$, one gets, from Eq. (4.59)

$$v^i = \frac{u^i(E) + v^i}{1 + \frac{v^i}{u^i(E)}} = u^i(E) \qquad (4.63)$$

whereas, for $v^{i\prime} = w^i(E)$, Eq. (4.61) yields

$$v^i = \frac{w^i(E) + v^i}{1 + \dfrac{(b_i(E))^2 v^i}{w^i(E)}} \neq w^i(E). \qquad (4.64)$$

We can therefore conclude, on a physical basis, that **u** is the maximal, invariant causal velocity in DSR, and it plays in the deformed Minkowski space $\tilde{M}$ the role of the light speed in standard SR.[d]

It is also easy to see why — although approach (II) looks at first sight more rigorous mathematically, because it permits to connect the peculiar features of spatial anisotropy of DSR to the deformed product $*$, "naturally induced" from the metric of $\tilde{M}(E)$ — actually it's approach (I) which yields the physically relevant result. Indeed, the velocity **u** is just defined as $\frac{d\mathbf{x}}{dt}$, and it therefore represents the physically measured velocity, for a particle moving in the usual, physical Euclidean 3-d space. On the other hand, this result clearly shows that the space anisotropy introduced by the deformed metric is not a mere mathematical artifact, but it reflects itself in the physical properties (imposed by the interaction involved) of the phenomenon described by the deformed space-time.

The comparison of the deformed boost expression (Eq. (4.9)) with the corresponding ones of the standard Lorentz boosts shows clearly that the transition from SR (based on M) to DSR (based on $\tilde{M}$) is simply carried out by letting

$$\mathbf{u}_{SR} = (c, c, c) \rightarrow \mathbf{u}_{DSR}(E) = \left(\frac{cb_0(E)}{b_1(E)}, \frac{cb_0(E)}{b_2(E)}, \frac{cb_0(E)}{b_3(E)} \right). \qquad (4.65)$$

In other words, the difference between M and $\tilde{M}(E)$ (at least as far as the finite coordinate transformations are concerned) is completely embodied in the 3-vector m.c.v. **u**.

---

[d]Of course, in the case of space isotropy, we get an isotropic maximal causal velocity given by (cf. Eq. (2.9))

$$u^i_{\text{iso}}(E) = u^i_{DSR,II}(E)|_{b_i(E)=b(E)} = c\frac{b_0(E)}{b(E)} \quad \forall\, i = 1, 2, 3$$

$$|\mathbf{u}_{\text{iso}}(E)| = \left(\sum_{i=1}^{3} (u^i_{\text{iso}}(E))^2 \right)^{1/2} = \sqrt{3}c\frac{b_0(E)}{b(E)}$$

4.4.4. *Choosing the Boost Direction in DSR*

We want now to remark a difficulty arising in the context of DSR, due to the space anisotropy.

Indeed, the space anisotropy (reflected in the physical anisotropic m.c.v. **u**) produces a triple indetermination in the process of identifying the motion axis with any of the space coordinate axes, since now — unlike the SR case — the space dimensions are no longer equivalent.

However, this indeterminacy can be removed (at least in principle) by means of the following *Gedankenexperiment*. Consider three particles (ruled by one and the same interaction) in general different but able to move at the maximal causal velocity $u^i(E)$. Suppose they are moving in the 3-d Euclidean space along mutually independent (orthogonal) spatial directions. Assigning an arbitrary labelling to the particle motion directions, we can fix an orthogonal, left-handed frame of axes. Since by assumption we know the interaction which the particles are subjected to, we know the deformed metric and therefore the metric coefficients as functions of the energy, $b_\mu^2(E)$. Then, a measurement of the particle velocities allows us to determine the right labelling of the spatial frame.

This implies that in the context of DSR, too, it is always possible, at physical level, to let one of the three space axes to coincide with the direction of motion of a physical object, and therefore apply the suitable deformed boost.

4.4.5. *Appendix — Another Derivation of a Generic Boost*

The procedure followed in Sec. 4.2 in order to derive the expression of a boost along a coordinate axis can in principle be exploited too in deriving the deformed boost in a generic direction. In this case, the coordinate transformations are

$$\begin{cases} x' = A_{11}x + A_{12}y + A_{13}z + A_{14}t \\ y' = A_{21}x + A_{22}y + A_{23}z + A_{24}t \\ z' = A_{31}x + A_{32}y + A_{33}z + A_{34}t \\ t' = A_{41}x + A_{42}y + A_{43}z + A_{44}t. \end{cases} \tag{A.1}$$

From the physical requirement that the origin O' of TIF K' must move in K with velocity components v^1 along $\hat{x}$, v^2 along $\hat{y}$, v^3 along $\hat{z}$, one gets:

$$\left. \begin{aligned} x' = 0,\ x = v^1 t \\ y' = 0,\ y = v^2 t \\ z' = 0,\ z = v^3 t \end{aligned} \right\} \Leftrightarrow \begin{cases} A_{11}v^1 + A_{12}v^2 + A_{13}v^3 + A_{14} = 0 \\ A_{21}v^1 + A_{21}v^2 + A_{31}v^3 + A_{24} = 0 \\ A_{31}v^1 + A_{32}v^2 + A_{33}v^3 + A_{34} = 0 \end{cases} . \tag{A.2}$$

Equation (A.1) becomes therefore

$$
\begin{cases}
x' = A_{11}(x - v^1 t) + A_{12}(y - v^2 t) + A_{13}(z - v^3 t) \\
y' = A_{21}(x - v^1 t) + A_{22}(y - v^2 t) + A_{23}(z - v^3 t) \\
z' = A_{31}(x - v^1 t) + A_{32}(y - v^2 t) + A_{33}(z - v^3 t) \\
t' = A_{41}x + A_{42}y + A_{43}z + A_{44}t.
\end{cases}
\tag{A.3}
$$

Replacing (A.3) in (A.1) yields

$$
\begin{aligned}
b_0^2(E) & c^2 t^2 - b_1^2(E)x^2 - b_2^2(E)y^2 - b_3^2(E)z^2 \\
&= c^2 b_0^2(E)(A_{41}x + A_{42}y + A_{43}z + A_{44}t)^2 \\
&\quad - b_1^2(E)(A_{11}(x - v^1 t) + A_{12}(y - v^2 t) + A_{13}(z - v^3 t))^2 \\
&\quad - b_2^2(E)(A_{21}(x - v^1 t) + A_{22}(y - v^2 t) + A_{23}(z - v^3 t))^2 \\
&\quad - b_3^2(E)(A_{31}(x - v^1 t) + A_{32}(y - v^2 t) + A_{33}(z - v^3 t))^2.
\end{aligned}
\tag{A.4}
$$

Equating the coefficients on both sides of (A.4) one gets a system of 10 quadratic equations in the 13 unknown coefficients $\{A_{ij}, A_{4i}\}$ $(i, j = 1, 2, 3)$, namely:

I. From the coefficient of t^2:

$$
c^2(A_{44}^2 - 1) - \frac{1}{b_0^2(E)} \sum_{i,j,l=1}^{3} b_j^2(E)v^j v^l A_{ij} A_{il} = 0.
\tag{A.5}
$$

II. From the coefficients of $x^i x^j$ $(i, j = 1, 2, 3)$, 6 independent equations:

$$
c^2 A_{4i} A_{4j} - \frac{1}{b_0^2(E)} \sum_{l=1}^{3} b_l^2(E)(A_{li} A_{lj} - \delta_{ij}\delta_{il}) = 0.
\tag{A.6}
$$

III. From the coefficients of $x^i t$ $(i = 1, 2, 3)$, 3 independent equations:

$$
c^2 A_{4i} A_{44} + \frac{1}{b_0^2(E)} \sum_{j,l=1}^{3} b_j^2(E)v^l A_{ji} A_{jl} = 0.
\tag{A.7}
$$

Although the above system in the set $\{A_{ij}, A_{4i}\}$ $(i, j = 1, 2, 3)$ can in principle be solved, the general solution for the boost expressed in the form (A.3) is quite cumbersome. This motivates the choice (adopted in Subsec. 4.4.1) of deriving the form of the deformed boost in a generic direction by exploiting the notion of "deformed" parallelism between 3-vectors.

CHAPTER 5

RELATIVISTIC KINEMATICS IN A DEFORMED MINKOWSKI SPACE

From the knowledge of the generalized Lorentz transformations it is easy to derive the main kinematical and dynamical laws valid in DSR.[3,4] In this Chapter, we shall merely list those which are useful to phenomenological purposes.

- **Velocity composition law** (cf. Eq. (4.58)):

$$V_{\text{tot}} = \frac{v_1 + v_2}{1 + \dfrac{v_1 v_2}{u^2}} \tag{5.1}$$

which obviously for, say, $v_1 = u$ yields $V_{\text{tot}} = u$.

If the condition of spatial isotropy is given up, the composition law for motion, say, along the x_k-axis, becomes

$$V = \frac{v_1 + v_2}{1 + \dfrac{v_1 v_2}{u_k^2}}; \quad u_k = \frac{c b_0}{b_k} \tag{5.2}$$

and, therefore, the speed that has an invariant character is

$$u_k = \frac{c b_0}{b_k}. \tag{5.3}$$

It follows that, in a given Minkowski space with deformed metric, there exist infinitely many different, maximal causal speeds, corresponding to the different possible directions of motion (although, of course, only three of them are independent). Clearly, this result is a strict consequence of the spatial anisotropy of the space-time region considered. Let us notice that there is indeed a phenomenon — the Bose–Einstein correlation — which can be fully described in the framework of such a Minkowski space, but with the consequence of a local loss of space isotropy (see Part III).

- **Time dilation**:

$$\Delta t = \tilde{\gamma}(E) \Delta t_0; \tag{5.4}$$

● **Length contraction**:

$$\Delta L = \tilde{\gamma}^{-1}(E)\Delta L_0; \tag{5.5}$$

● **Four-velocity**:

$$V^\mu = \frac{d}{dt_0}x^\mu, \tag{5.6}$$

whose explicit form (with dt_0 derived from Eq. (5.4)) reads

$$V^0(E) = \tilde{\gamma}(E)u(E); \tag{5.7}$$
$$V^k(E) = \tilde{\gamma}(E)u^k(E). \tag{5.8}$$

Therefore, the generalized expression of the momentum fourvector (in the case of spatial isotropy) is

$$p^\mu = m_0 V^\mu(E) = (m_0\tilde{\gamma}(E)u(E), m_0\tilde{\gamma}(E)v^k). \tag{5.9}$$

In the general case, the deformed relativistic energy, for a particle subjected to a given interaction and moving along $\hat{x}^i$, has the form:

$$E = mu_i^2(E)\tilde{\gamma}(E) = mc^2\frac{b_0^2(E)}{b_i^2(E)}\tilde{\gamma}(E) \tag{5.10}$$

where $\mathbf{u}(E)$ is the maximal causal velocity (4.31) for the interaction considered. In the non-relativistic (NR) limit of DSR, *i.e.* at energies such that

$$v_i \ll u_i(E). \tag{5.11}$$

Equation (5.10) yields the following NR expression of the energy corresponding to the given interaction:

$$E_{NR} = mu_i^2(E) = mc^2\frac{b_0^2(E)}{b_i^2(E)}. \tag{5.12}$$

Lastly, let us consider a plane wave propagating with speed u (e.g. in the xy plane, at angles θ, θ' in frames K, K') with dispersion relation $u = \lambda\nu = \lambda'\nu'$, where ν, ν' are the wave frequencies in K, K'. Applying the generalized Lorentz transformations, it is easy to get the following laws:

● **Doppler effect**:

$$\nu = \tilde{\gamma}(E)\nu'(1 + \tilde{\beta}(E)\cos\theta'); \tag{5.13}$$

● **Aberration law:**

$$tg\theta = \frac{\sin\theta'}{\tilde{\gamma}(E)(\tilde{\beta}(E) + \cos\theta')}. \tag{5.14}$$

We want now to provide a comparison between the main kinematical laws in the usual Minkowski space M and in the deformed one $\tilde{M}$ (in the hypothesis of spatial isotropy), because their different behaviors may help one to understand the peculiar features of leptonic, hadronic (and gravitational) interactions with respect to the electromagnetic one. Such laws are summed up in Table 5.1, where the maximal speed u has been expressed in terms of c, in order to emphasize the dependence of the deformed laws on the parameter ratio b/b_0 and exhibit their scale invariance.

In the limiting case $v = c$, one gets explicitly

$$v_1 = c \Rightarrow V_{\text{tot}} = \frac{c + v_2}{1 + \left(\dfrac{b}{b_0}\right)^2 \dfrac{v_2}{c}}; \tag{5.15}$$

$$v = c \Rightarrow \Delta t = \frac{\Delta t_0}{\left[1 - \left(\dfrac{b}{b_0}\right)^2\right]^{1/2}}; \tag{5.16}$$

$$v = c \Rightarrow \Delta L = \Delta L_0 \left[1 - \left(\dfrac{b}{b_0}\right)^2\right]^{1/2}. \tag{5.17}$$

Remember that, in this framework, c has lost its meaning of maximal causal speed, by preserving the mere role of maximal causal speed for electromagnetic phenomena in M.

Table 5.1

Minkowski space	Deformed Minkowski space
$V_{\text{tot}} = \dfrac{v_1 + v_2}{1 + \dfrac{v_1 v_2}{c^2}}$	$V_{\text{tot}} = \dfrac{v_1 + v_2}{1 + \left(\dfrac{b}{b_0}\right)^2 \dfrac{v_1 v_2}{c^2}}$
$\Delta t = \dfrac{\Delta t_0}{\left(1 - \dfrac{v^2}{c^2}\right)^{1/2}}$	$\Delta t = \dfrac{\Delta t_0}{\left[1 - \left(\dfrac{b}{b_0}\right)^2 \dfrac{v^2}{c^2}\right]^{1/2}}$
$\Delta L = \Delta L_0 \left(1 - \dfrac{v^2}{c^2}\right)^{1/2}$	$\Delta L = \Delta L_0 \left[1 - \left(\dfrac{b}{b_0}\right)^2 \dfrac{v^2}{c^2}\right]^{1/2}$

Table 5.2

Minkowski space	Deformed Minkowski space
$\Delta t = \Delta t_0 \dfrac{E}{m_0}$	$\Delta t = \Delta t_0 \left[1 - \left(\dfrac{b}{b_o} \right)^2 + \left(\dfrac{b}{b_o} \right)^2 \left(\dfrac{m_0}{E} \right)^2 \right]^{-1/2}$
$\Delta L = \Delta L_0 \dfrac{m_0}{E}$	$\Delta L = \Delta L_0 \left[1 - \left(\dfrac{b}{b_o} \right)^2 + \left(\dfrac{b}{b_o} \right)^2 \left(\dfrac{m_0}{E} \right)^2 \right]^{1/2}$

To the purpose of an experimental verification, it is worth to express the deformed kinematical laws of time dilation and length contraction for a particle of rest mass m_0 in terms of the *usual* energy E. Clearly, for $E \gg m_0 c^2$, E can be considered the total energy of the particle, measured (as already stressed in Sec. 3.2) by electromagnetic methods in the usual Minkowski space. We report such laws in Table 5.2 (in comparison with the standard, Einsteinian ones).

It is easily seen that, in the case of the time-dilation law, the main difference is the loss of linearity in the dependence on the energy of the deformed law, as compared to the Lorentzian one. Such a different behaviour is therefore a clear signature of the presence of nonlocal effects in the interaction considered. A first evidence is provided by the lifetime of the meson K_s^0 in the range $3 \div 400\,\mathrm{GeV}$, as it will be seen in Part III.

CHAPTER 6

WAVE PROPAGATION IN A DEFORMED SPACE-TIME

6.1. Deformed Helmholtz Equation

We want now to approach the problem of wave propagation in a deformed Minkowski space-time.[13,14] To this end, let us introduce the generalized D'Alembert operator $\tilde{\Box}$, defined by means of the scalar product $*$ in $\tilde{M}$ (see Eq. (2.5)):

$$\tilde{\Box} \equiv \partial * \partial = \eta_{\mu\nu}\partial^\mu\partial^\nu = \frac{b_0^2}{c^2}\partial_t^2 - (b_1^2\partial_x^2 + b_2^2\partial_y^2 + b_3^2\partial_z^2). \tag{6.1}$$

Therefore, the deformed Helmholtz–D'Alembert wave equation is given by

$$\tilde{\Box}f = 0 \tag{6.2}$$

with f being any component of the field associated to the wave considered. For instance, the field of such a wave propagating in the Minkowski space $\tilde{M}$ can be written as

$$\mathbf{f}(x) = \mathbf{A}(\mathbf{x})e^{ik*x} \tag{6.3}$$

where k is the wavevector and e^{ik*x} is the generalized phase.

By assuming a spatially isotropic deformed metric (see Eq. (2.7)), in the corresponding deformed space-time the generalized phase takes the "Minkowskian-like" form $e^{i\tilde{k}\cdot x}$ (where the dot denotes the usual scalar product in the Minkowski space), with

$$\tilde{k}^\mu = \left(\frac{2\pi\nu}{c}, \tilde{k}_x, \tilde{k}_y, \tilde{k}_z\right) \tag{6.4}$$

and ν is the frequency measured in the ordinary space-time. Then, Eq. (6.3) becomes

$$\mathbf{f}(x) = \mathbf{A}(\mathbf{x})e^{i\tilde{k}\cdot x}. \tag{6.5}$$

A wave propagating in a deformed Minkowski space-time will be referred to in the following as a *non-Lorentzian wave*.

6.2. Propagation in an Undersized Waveguide

As an application of the previous results leading to interesting physical consequences, we want now to discuss the case of an electromagnetic (e.m.) field propagating in a conducting waveguide, with reduced section of width a (see Fig. 6.1), in the principal transversal electric mode (commonly denoted TE_{10}). If the guide axis coincides with the z-axis, we can put $E_x = E_z = 0$, and the only nonvanishing component E_y can be written as

$$E_y(x, z, t) = g(x)e^{i(\tilde{k}_z z - \omega t)}. \tag{6.6}$$

Therefore the generalized Helmholtz equation (6.2) for E_y becomes

$$b^2 \partial_x^2 E_y + b^2 \partial_z^2 E_y - \frac{b_0^2}{c^2} \partial_t^2 E_y = 0 \tag{6.7}$$

or

$$b^2 \partial_x^2 g - b^2 \tilde{k}_z^2 g + \frac{b_0^2}{c^2}(2\pi\nu)^2 g = 0. \tag{6.8}$$

This last equation for $g(x)$ has solution

$$g(x) = C_1 \cos(\tilde{k}_x x) + C_2 \sin(\tilde{k}_x x) \tag{6.9}$$

with C_1, C_2 constants. Imposing the boundary conditions on E_y:

$$E_y = 0, \quad x = 0, a \tag{6.10}$$

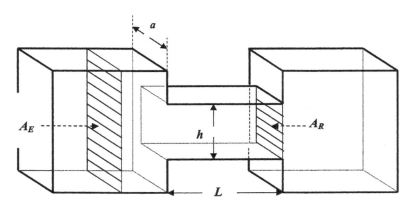

Fig. 6.1. Rectangular waveguide with variable section used in the Cologne experiment. L = length of the narrow part of the waveguide ("barrier"); h = height of the guide; a = tickness of the guide; A_E = area of the large section ("emitter"); A_R = area of the small section ("receiver").

it is obviously

$$g(x) = 0, \quad x = 0, a \tag{6.11}$$

so that $C_1 = 0, C_2 \neq 0$ and

$$\tilde{k}_x = \frac{\pi}{a} m, \quad m = 1, 2, \ldots \tag{6.12}$$

Replacing (6.9) and (6.12) in Eq. (6.8), since $m = 1$ in the TE_{10} mode, one gets

$$-b^2 \left(\frac{\pi}{a}\right)^2 - b^2 \tilde{k}_z^2 + b_0^2 \left(\frac{2\pi}{c}\right)^2 \nu^2 = 0. \tag{6.13}$$

Such a relation yields the wavevector component $\tilde{k}_z$ along the propagation direction in a deformed Minkowski space with metric (2.7). Putting

$$\frac{\pi}{a} = \frac{2\pi}{c} \nu_c, \tag{6.14}$$

where ν_c is the cutoff frequency of the waveguide, one finds from (6.13) the explicit expression of $\tilde{k}_z^2$:

$$\tilde{k}_z^2 = \left(\frac{2\pi}{c}\right)^2 \nu^2 \left[\frac{b_0^2}{b^2} - \left(\frac{\nu_c}{\nu}\right)^2\right]. \tag{6.15}$$

In the case of a Minkowskian metric ($b_0^2 = b^2 = 1$), Eq. (6.15) is nothing but the usual relation for guided e.m. waves. The condition of actual propagation is obviously given by $\nu > \nu_c$ (high-pass filter), whereas for $\nu < \nu_c$ there is no propagation in a strict sense, but only an evanescent mode with $\tilde{k}_z^2 < 0$ (imaginary wavevector).

On the contrary, in the case of a deformed space-time, the condition of actual propagation depends in an essential way on the ratio b_0^2/b^2. Indeed, even for $\nu < \nu_c$ it is possible to have $\tilde{k}_z^2 > 0$, provided that relation

$$\frac{b_0^2}{b^2} > \left(\frac{\nu_c}{\nu}\right)^2 \tag{6.16}$$

holds.

This defines the validity conditions of (6.7) and then of (6.15). Moreover, when $(\nu_c/\nu) > 1$, it follows $(b_0/b) > 1$, namely, by Eq. (2.10), $u > c$. Relation (6.16) can be therefore regarded as the condition of propagation for a wave with *real* wavevector, and *superluminal speed*, in the deformed Minkowski space, in correspondence to frequency values ($\nu < \nu_c$) for which, in the usual Minkowski space, the wavevector is imaginary, and the related e.m. wave is therefore in an evanescent mode. Otherwise speaking, for suitable values of the ratio b_0/b, and in suitable space-time regions, an evanescent e.m. wave in the usual space can be interpreted as an actual wave propagating, with superluminal speed, in the deformed Minkowski space.

Such a (classical) interpretation of an imaginary wavevector can be straightforwardly extended to the quantum case (which amounts therefore to describing virtual photons as tachyonic "photons" with speed $u > c$). Let us notice that, in this framework, the wave speed is already fixed by metric (2.2), and *not* by the Helmholtz equation (6.7), which yields only the propagation conditions in a space with generalized metric η and maximal causal speed u (which, as stressed in Chap. 2, is the speed of the "photons" associated to the interaction described by the given metric).

Moreover, as already stressed in Sec. 3.2, these tachyonic photons — massless with respect to the interaction described by the deformed metric, and therefore in the corresponding deformed Minkowski space $\tilde{M}$ — could appear as massive when observed in the usual Minkowski space M (*i.e.* viewed by the e.m. interaction), where they however propagate as evanescent waves.

6.3. Total Internal Reflection

As another application, let us consider the case of total internal reflection of light.[15] Consider two right-angle glass prisms, with their faces corresponding to hypothenuse separated by a thin air gap of width L, and a light wave incident on the first prism at an angle $i > i_c$ (where $i_c = \sin^{-1}\left(\frac{1}{n}\right)$ is the critical angle of total reflection, if $n > 1$ is the refraction index of the glass). Then, the wave is transmitted in the second prism, beyond the interface gap, only in the form of an exponentially attenuated (evanescent) wave.

The schematic picture of the optical device used in these kind of experiments[16] is given in Fig. 6.2, where the adopted notation remarks the analogy with the e.m. case (see Fig. 6.1). The following symbols are used: L = length of the slab; i = incidence angle; D = spatial shift; $\tau = \tau_\phi$ = average temporal shift. We assume that the behaviour of the air slab between the glasses is *nonlocal* (as far as the light propagation is concerned), so that the space-time inside the slab is endowed with the deformed metric

$$ds^2 = b_0^2(\mathcal{E})c^2 dt^2 - b_1^2(\mathcal{E})dx^2 - b_2^2(\mathcal{E})dy^2 - b_3^2(\mathcal{E})dz^2 = \eta_{\mu\nu}(\mathcal{E})dx^\mu dx^\nu;$$
$$\eta_{\mu\nu}(\mathcal{E}) = \mathrm{diag}\left(b_0^2(\mathcal{E}), -b_1^2(\mathcal{E}), -b_2^2(\mathcal{E}), -b_3^2(\mathcal{E})\right) \tag{6.17}$$

where $\mathcal{E}$ is the energy of the light beam.

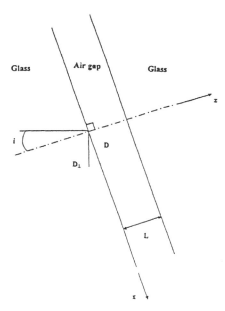

Fig. 6.2. Schematic view of the optical tunneling device used in the experiment. The glass (dielectric medium) has refraction index n. The light beam impinges onto the air gap of width L at an incidence angle i, and undergoes both a spatial shift D (along x) and a temporal one τ. It is $D_\perp = D/\sin i$.

If n is the refraction index of the glass, the spatial shift D is given by[16]

$$D = \tau c \frac{\cos i}{n} + L \frac{\sqrt{2}}{2} \cos i \sin i. \tag{6.18}$$

Then, the maximal causal speeds of light inside the slab along the directions x and z are expressed in terms of the deformation coefficients as

$$u_x(\mathcal{E}) = \frac{b_0(\mathcal{E})}{b_1(\mathcal{E})} c; \quad u_z(\mathcal{E}) = \frac{b_0(\mathcal{E})}{b_3(\mathcal{E})} c \tag{6.19}$$

and they are functions of the beam energy $\mathcal{E}$. On the other hand, it is

$$u_x(\mathcal{E}) = \frac{D}{\tau}; \quad u_z(\mathcal{E}) = \frac{L}{\tau}. \tag{6.20}$$

Therefore, from the above two relations, one gets

$$\frac{b_3(\mathcal{E})}{b_1(\mathcal{E})} = \frac{u_x(\mathcal{E})}{u_z(\mathcal{E})} = \frac{D}{L} = \frac{\tau c \cos i}{L} \frac{1}{n} + \frac{\sqrt{2}}{2} \cos i \sin i \tag{6.21}$$

which expresses the ratio of the metric coefficients in terms of the optical device parameters and of the measured time shift.

In order to find the time metric coefficient b_0, we need an expression of the energy of the light beam as a function of z. The heuristic assumption is

$$\mathcal{E} = \mathcal{E}_0 e^{-z/z_0}. \tag{6.22}$$

Here, $\mathcal{E}_0$ is the initial energy of the light beam, which for a monochromatic beam of $\mathcal{N}$ photons with frequency ν bounded in space and time (of duration T) reads

$$\mathcal{E}_0 = \mathcal{N}h\nu \tag{6.23}$$

and z_0 is the cutoff width of the slab (see below).

From Eqs. (6.22) and (6.23) we get

$$\mathcal{E} = \mathcal{N}h\nu e^{-z/z_0} = \mathcal{N}h\nu' \tag{6.24}$$

where ν' is to be understood as the frequency at distance z inside the slab. On the other hand, it is

$$\nu' = \frac{1}{b_0 T} \tag{6.25}$$

and therefore the time coefficient has the form

$$b_0 = e^{z/z_0}. \tag{6.26}$$

The energy lost by the beam as it travels a distance z inside the slab is

$$\Delta\mathcal{E} = \mathcal{E}_0 - \mathcal{E} = \mathcal{N}h\nu \left(1 - e^{-z/z_0}\right) \tag{6.27}$$

and it can be considered as the energy needed to deforming the space-time inside the slab.

By analogy with the case of the waveguide, where the cutoff length L_0 is given by

$$L_0 = \frac{\lambda}{2\pi} \left[\left(\frac{\lambda}{\lambda_c}\right)^2 - 1\right]^{-1/2} \tag{6.28}$$

(where λ_c is the critical wavelength) we can put, in the optical case

$$z_0 = \frac{\lambda}{2\pi} \left[\left(\frac{\sin i}{\sin i_c}\right)^2 - 1\right]^{-1/2} \tag{6.29}$$

with i_c and λ being, respectively, the critical angle and the wavelength of the monochromatic light beam in the glass. This is essentially due to the tunneling interpretation of both phenomena and the analogous role played by the critical wavelength and the critical angle in the two cases.

On the basis of the above analogy, we can also write the following expression for the wavevector along the z-direction inside the slab:

$$\tilde{k}_z^2 = \left(\frac{2\pi}{\lambda}\right)^2 \left[\left(\frac{b_0}{b_3}\right)^2 - \left(\frac{\sin i}{\sin i_c}\right)^2\right]. \tag{6.30}$$

In the case $i_c < i$, which classically (in Minkowskian conditions) corresponds to an imaginary wavevector (in the slab deformed space-time), one gets instead a real wavevector ($\tilde{k}_z^2 > 0$), provided that the following reality condition holds:

$$\left(\frac{b_0}{b_3}\right)^2 > \left(\frac{\sin i}{\sin i_c}\right)^2 > 1 \tag{6.31}$$

and this case yields a superluminal speed of the light beam inside the slab, according to Eq. (2.10):

$$u_z = \frac{b_0}{b_3}c > c. \tag{6.32}$$

This is why condition (6.31) is also referred to as the *superluminality condition* (cf. (6.16)).

By the formalism of the deformed Minkowski space-time, it is possible also to give an expression of the time-averaged Poynting vector in the z-direction for a monochromatic component of the light beam. One has

$$\langle S_z \rangle = \left(\frac{c^2}{2\pi\nu}\right)\tilde{k}_z \langle U \rangle_\ell \tag{6.33}$$

where $\langle U \rangle_\ell$ is the energy density averaged on ℓ, with ℓ varying in the range $(0, L\sin i)$, given by

$$\langle U \rangle_\ell = \frac{1}{16\pi}E_{0x}^2 \tag{6.34}$$

(with E_{0x} being the peak value of the electric field of the light beam). On account of Eq. (6.33), one finds therefore

$$\langle S_z \rangle = \frac{c}{16\pi^2}\left[\left(\frac{b_0}{b_3}\right)^2 - \left(\frac{\sin i}{\sin i_c}\right)^2\right]^{1/2}E_{0x}^2. \tag{6.35}$$

6.4. Deformation Tensor

In both the examples considered above, the evanescent wave inside the barrier can be expanded in a Fourier series as[14]

$$e^{-\chi z} = \sum_{n=-\infty}^{\infty} c_n e^{in2\pi z/L}, \tag{6.36}$$

where

$$\chi = \frac{2\pi}{c}\sqrt{\nu_c^2 - \nu^2} \tag{6.37}$$

and (for $\chi L \gg 1$)

$$c_n = \frac{(1/L)}{\chi + in\dfrac{2\pi}{L}}. \tag{6.38}$$

Notice that the relations (6.15)–(6.16) hold for each Fourier component of the expansion (6.36). As a consequence, all the related propagating waves in (6.36) are superluminal. Moreover, each Fourier component propagates in a different deformed Minkowski space-time. This is clearly related to the energy (and momentum) dependence of the parameters of the deformed metric. If $\eta_{\mu\nu}^{(n)}$ is the deformed metric "seen" by the nth Fourier component of the evanescent wave, with metric coefficients $b_\mu^{(n)}$, we can build an *effective metric tensor* $\bar{\eta}_{\mu\nu}$ for the evanescent wave as follows[a]:

$$\bar{\eta}_{\mu\nu}(c_n) = \frac{\sum_n |c_n|^2 \eta_{\mu\nu}^{(n)}}{\sum_n |c_n|^2}. \tag{6.39}$$

Clearly, outside the barrier, all the Fourier waves propagate in a Minkowskian space-time, and definition (6.39) reduces to the usual metric $g_{\mu\nu}$. In fact, in such a region we are in full Minkowskian conditions (since the energy of the process is higher than the electromagnetic threshold energy, $E > E_{0\text{e.m.}}$), *i.e.* $b_\mu^{(n)2} = 1$, $\mu = 0, 1, 2, 3$, and therefore $\eta_{\mu\nu}^{(n)} = g_{\mu\nu} \ \forall n$.

Notice that the tensor (6.39) is analogous to the Cauchy stress tensor of a continuous medium. In fact, let us consider, in orthogonal Cartesian coordinates, an infinitesimal tetrahedron with edges parallel to the coordinate axes and the oblique face S opposite to the vertex O, origin of the Cartesian frame. If the tetrahedron is a part of a continuous body, the stress vector

---

[a]Analogous results hold in the case of a growing wave, or when both an evanescent and a growing wave are present inside the barrier.

across S in the point O is given by

$$\phi_{\mathbf{a}}(O) = \frac{\sum_i \phi_{\mathbf{i}}(O)a_i}{\sum_i |a_i|^2} \tag{6.40}$$

where $\mathbf{a}$ is a vector normal to S and $\phi_i(O)$ $(i = 1, 2, 3)$ is the stress vector on the face of the tetrahedron orthogonal to the ith axis. The nine components of the three vectors $\phi_i(O)$ do just constitute the rank-two, symmetric Cauchy tensor.

The tensor $\bar{\eta}$ can be therefore regarded as the average tensor representing the space-time deformation inside the barrier (corresponding to the energy $E < E_{0\text{e.m.}} \sim 5\,\mu\text{eV}$: see Chap. 7) globally "seen" by the evanescent wave (6.36). So, we can name it *average tensor of the electromagnetic space-time deformation*, $\bar{\eta}_{\text{e.m.}}$.

It is worth stressing that such an approach to electromagnetic faster-than-light propagation is similar, in some respects, to that where superluminal propagation (e.g. of light between parallel mirrors) is connected to vacuum effects.[17] In this case, the influence of the (structured) vacuum is described in an effective way in terms of a refractive index (as pioneered by Sommerfeld). Something analogous happens in General Relativity, too: the deflection of light rays in a gravitational field can be considered as a propagation in an Euclidean space, filled with a medium endowed with an effective refraction index.[18] In some cases, such a propagation — due to the influence of the gravitational vacuum — turns out to be superluminal (the refractive index is less than one).[19] The deformed metric approach can be therefore regarded as *dual* to the general relativistic one, in which the space-time curvature for electromagnetic signals is replaced by a refractive index. In the DSR formalism, the vacuum or nonlocal effects which affect propagation inside the barrier are described in terms of a space-time deformation (and the role of the refractive index is played by the deformation tensor).

Let us emphasize that the definition (6.39) of the deformation tensor can be applied also to non-Lorentzian wavepackets ruled by interactions different from the electromagnetic one. Namely, it can be stated in full generality that the Minkowski space is always subjected to a stress, whenever crossed by a wavepacket. Such a stress is related to the deformation of the space-time, which may be described by the tensor $\bar{\eta} = g$ (*ineffectual deformation*) or by a tensor $\bar{\eta} \neq g$ (*effectual deformation*). The two cases $\bar{\eta} = g$, $\bar{\eta} \neq g$ are obviously determined by the interaction ruling the wavepacket propagation and by the energy of the wavepacket components.

In conclusion, let us notice that the choice of the examples used to discuss the generalized Helmholtz–D'Alembert equation (6.2) is due to two reasons. First, this allowed us to provide concrete cases of metric description of electromagnetic interaction in non-Minkowskian conditions. Moreover, the condition $\nu < \nu_c$ in a waveguide (or $\lambda < \lambda_c$ in the optical case) corresponds, in quantum mechanics, to the crossing of a barrier potential by tunnel effect. The latter analogy suggests a possible (although highly speculative) interpretation of quantum tunneling as a wave propagation in a deformed space-time, whose metric is determined by the interaction generating the barrier, with coefficients b_μ evaluated at that energy value corresponding on average to the barrier height. Of course, such a metric would have to be restricted to the space region occupied by the barrier.

PART III

METRIC DESCRIPTION OF FUNDAMENTAL INTERACTIONS

CHAPTER 7

NONLOCAL EFFECTS IN ELECTROMAGNETIC INTERACTION

In this third Part the formalism of Deformed Special Relativity, developed in the first two Parts, will be applied to provide examples of the phenomenological description of interactions by means of deformed (topical) metrics, on the basis of the allowed experimental data on some electromagnetic, gravitational, leptonic and hadronic phenomena. The cases we shall discuss will illustrate the practical use of DSR, on one side, and yield some preliminary evidence on a possible breakdown of Lorentz invariance (and, therefore, on the actual presence of nonlocal effects in fundamental interactions).

7.1. Non-Minkowskian Metric for the E.M. Interaction

7.1.1. *Photon Tunneling and Superluminal Velocities*

Tunneling of a particle through a potential barrier is a well-known quantum effect, which finds a lot of applications ranging from the tunnel diode to the scanning tunneling microscope. Nevertheless, a lot of controversies still exist on the seemingly simple-sounding question of the time taken by a particle to tunnel.[20-22] We do not want to enter here in such a debate, and only confine ourselves to quote that most of the theoretical definitions of tunneling time do imply that the tunneling process occurs at faster-than-light speed. This is essentially related to the so-called *Hartman-Fletcher* (*HF*) *effect*[23,24]: the tunneling time is independent of the barrier width d for sufficiently large d. The HF effect has been proved to hold for all the mean (nonrelativistic) tunneling times.

A well-known optic analog of the quantum tunneling is provided by total internal reflection (discussed in Subsec. 6.3). This phenomenon is formally analogous to the (one-dimensional) tunneling through a potential barrier of height V_0 and width L by a particle with energy $\mathcal{E} < V_0$. The analogy is rooted in the formal identity between the classical Helmholtz equation describing electromagnetic wave propagation and the quantum

Schrödinger equation for a particle. The general form of such an equation can be written as

$$\nabla^2 f + \kappa^2(\mathbf{r})f = 0$$

where f is any component of the electromagnetic field for the electromagnetic case, or the wave function ψ in the quantum case, and the "wavevector" $\kappa(\mathbf{r})$ depends on the (classical or quantum) case and on the specific problem considered.

For a particle of mass m and energy $\mathcal{E}$ in a potential $V(\mathbf{r})$, the (time-independent) Schrödinger equation reads

$$\nabla^2 \psi + \frac{2m}{\hbar^2}[\mathcal{E} - V(\mathbf{r})]\psi = 0$$

so

$$\kappa = \frac{1}{\hbar}\sqrt{2m[\mathcal{E} - V(\mathbf{r})]}.$$

In particular, for a uniform potential barrier of height V_0, inside the barrier κ becomes imaginary for $\mathcal{E} < V_0$.

For a monochromatic wave of frequency ω in an inhomogeneous and isotropic medium of refraction index $n(\mathbf{r})$, κ is given by

$$\kappa(\mathbf{r}) = \frac{n(\mathbf{r})\omega}{c}$$

whereas, in the case of the two-prism system, one has, inside the air gap

$$\kappa = \frac{\omega}{c}\sqrt{1 - n^2\sin^2 i} = \frac{\omega}{c}\sqrt{1 - \frac{\sin^2 i}{\sin^2 i_c}}$$

$(i_c = \sin^{-1}\left(\frac{1}{n}\right))$.

Finally, for e.m. wave propagation in waveguides, it is

$$\kappa = \frac{n}{c}\sqrt{\omega^2 - \omega_c^2}$$

where ω_c is the cutoff frequency of the waveguide.

The above equations are at the very basis of *the particle-photon tunneling analogy*.[25–28] The condition $\mathcal{E} < V_0$ in the tunneling of a particle through a potential barrier corresponds to an incident angle $i > i_c$, for the total internal reflection, and to $\omega < \omega_c$, for an evanescent mode in a waveguide. Therefore, one expects on theoretical grounds that in such cases photon tunneling takes place at superluminal speed. This is exactly what has been observed since 1992 by some experimental groups.[16,29–41]

The observed speed is the *group velocity* v_g of the wavepacket, defined as usual by

$$v_g = \frac{d\omega}{dk}$$

(the phase velocity being given by $v_p = c/n$).

Superluminal photon tunneling has been observed in both the microwave range and in the optical domain in experiments performed at Cologne, Florence, Berkeley, Vienna, Orsay, Rennes.

7.1.2. *Deformed E.M. Metric from the Cologne Experiment*

The basic assumption in order to apply the DSR formalism to superluminal photon propagation is that the presence of superluminal e.m. speeds in such experiments is to be ascribed to some nonlocal effects in e.m. interactions, which admit of an effective description in terms of a space-time deformation, according to the results of Part I.

In order to derive the explicit form of the e.m. metric describing such nonlocal effects, we shall analyze the Cologne experiments[29-31] on the propagation of e.m. waves in conducting waveguides with variable section.

Let us apply, in particular, the content of Sec. 6.2. The scheme of the waveguide with variable section is shown in Fig. 6.1, whereas the speed of an evanescent mode in crossing the reduced section waveguide with length L is plotted, in Fig. 7.1, as a function of L.

Indeed, the wave frequency ν in the guide with greater section was lower than the cutoff frequency ν_c of the guide with smaller section. On account

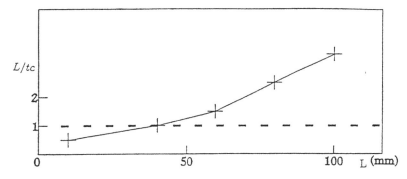

Fig. 7.1. Plot of the transmission velocity L/tc vs. the length L of the reduced waveguide (Cologne experiment).

of the mode involved (TE_{10}), it is possible to assume that space is isotropic inside the smaller waveguide. Moreover, since Eq. (3.8) depends only on the ratio (b/b_0), we can also put $b_0^2 = 1$ (*i.e.* the deformed metric is isochronous with the Minkowskian one). Therefore, the metric to be considered is

$$\eta = \text{diag}(1, -b^2, -b^2, -b^2). \qquad (7.1)$$

Equation (3.8) then becomes

$$u = \frac{c}{b} \qquad (7.2)$$

and this relation allows one to deduce the values of the parameter b. The wavevector is given by Eq. (6.15), that with metric (7.1) takes the form:

$$\tilde{k}_z^2 = \left(\frac{2\pi}{c}\right)^2 \nu^2 \left[\frac{1}{b^2} - \left(\frac{\nu_c}{\nu}\right)^2\right]. \qquad (7.3)$$

Equation (7.3) permits, in correspondence to the computed values of b, to find the values of the wavevector component $\tilde{k}_z$ for the frequencies used in the experiments, *i.e.* $\nu = 8.70\,\text{GHz}$ and $\nu_c = 9.49\,\text{GHz}$.

The ratio ν_c/ν yields, according to (6.16), the condition whereby $\tilde{k}_z^2 \geq 0$ holds, *i.e.*

$$\frac{1}{b^2} > \left(\frac{\nu_c}{\nu}\right)^2 = 1.091. \qquad (7.4)$$

Moreover, exploiting the classical theory of an evanescent mode, it is possible, from the knowledge of the length L of the reduced-section guide, to evaluate the output energies corresponding to given values of L and b. The exploited relation is (according to the heuristic assumption of Sec. 6.3):

$$E = h\nu e^{-L/L_0} \qquad (7.5)$$

where h is the Planck constant, ν the photon frequency at the input of the reduced-section guide, and L_0 is the penetration length *of the energy*. Notice that Eq. (7.5) does not mean at all that fractions of the photon energy are involved. As for the case of total internal reflection, the meaning of the above expression is simply that the energy lost by the signal after travelling a distance L inside the reduced waveguide

$$\Delta E = E_{\text{in}} - E = h\nu \left(1 - e^{-L/L_0}\right) \qquad (7.6)$$

can be considered as the energy needed to deforming the space-time inside the barrier.

The length L_0 is just provided by the classical theory of the evanescent mode:

$$L_0 = \left(\frac{c}{2\pi}\right) \frac{1}{\sqrt{\nu_c^2 - \nu^2}}. \tag{7.7}$$

For the frequency values used in the experiments, L_0 is real because $\nu < \nu_c$, and it is $L_0 = 1.256\,\text{cm}$.

Figure 7.2 shows the behaviour of the parameter b^2 vs. E for values $u > c$. The interpolation curve has been obtained using the values of E and b^2 for which $\tilde{k}_z^2 > 0$. We give below the explicit functional form of the corresponding metric $\eta_{\text{e.m.}}(E)$, that is to be regarded as a mere phenomenological description of the e.m. interaction under nonlocal conditions[4,13]:

$$\eta_{\text{e.m.}}(E) = \text{diag}(1, -b_{\text{e.m.}}^2(E), -b_{\text{e.m.}}^2(E), -b_{\text{e.m.}}^2(E))$$

$$b_{\text{e.m.}}^2(E) = \begin{cases} (E/E_{0,\text{e.m.}})^{1/3}, & 0 \le E < E_{0,\text{e.m.}} = 4.5 \pm 0.2\,\mu\,\text{eV} \\ 1, & E_{0,\text{e.m.}} \le E \end{cases} \tag{7.8}$$

The threshold energy $E_{0,\text{e.m.}}$ is the energy value at which the metric parameters are constant, *i.e.* the metric becomes Minkowskian, and the electromagnetic interaction is fully derivable from a potential. Notice that $\eta_{\text{e.m.}}(E)$ is a *subminkowskian* metric, namely it attains the Minkowskian limit from below (for $E < E_{0,\text{e.m.}}$).

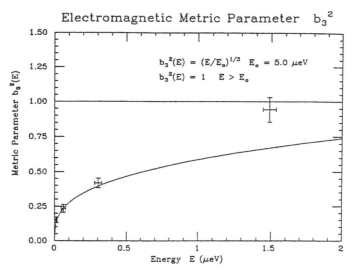

Fig. 7.2. Plot of the electromagnetic metric parameter $b^2(E)$ vs. the energy of the output signal.

Using the explicit form of $\eta_{\text{e.m.}}(E)$, Eq. (7.8), it is also possible to find the maximal causal speed $u_{\text{e.m.}}$, in the electromagnetic case, as function of the energy. From Eqs. (7.2) and (7.8), one gets

$$u_{\text{e.m.}}(E) = \begin{cases} (E_{0,\text{e.m.}}/E)^{1/6}, & 0 \leq E \leq E_{0,\text{e.m.}} = 4.5 \pm 0.2\,\mu\,\text{eV} \\ 1, & E_{0,\text{e.m.}} < E \end{cases} \qquad (7.9)$$

(in units of c). Figure 7.3 shows the behaviour of the maximal causal speed (7.9).

For a check, the results of such a metric description of electromagnetic interaction have been applied to the crossing of a translucid mirror by laser photons with wavelength 702 nm, which, in interferometric measurements, yield the value $u = 1.7c$ for the photon speed. To this aim, one needs to exploit the analogy between a translucid mirror crossing and the crossing of a conducting guide, on account of the fact that the aluminium-platings are alternate conducting and dielectric layers. One gets, for the transmitting part of the mirror, a diameter of about $0.5\,\mu$m and a length of about $2\,\mu$m, which are typical values for the aluminium-platings of such mirrors (frequently used in resonant cavities), as in the Berkeley experiment. Such a check shows the consistency of DSR in analyzing these two different experiments.

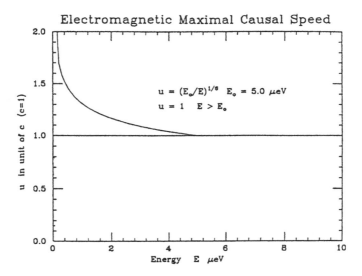

Fig. 7.3.   Plot of the electromagnetic maximal causal speed vs. energy (in units of c).

To end this section, let us briefly comment on the behaviour of the e.m. metric parameter, deduced from the experiments on superluminal photon tunneling.

Firstly, it is seen from Fig. 7.2 that the departure from the standard Minkowskian metric occurs at about $4.5 \div 5\,\mu eV$. This is just the energy corresponding to the "coherence wavelength" λ of the photon, defined by $E_{\text{coh}} = hc/\lambda$. Indeed, in the case of radio-optical sources, with frequencies of the order 10^{-8} sec, and $\lambda \approx 10^{2}$ cm, it is $E_{\text{coh}} \simeq 1\,\mu eV$, *i.e.* the coherence energy is just of the order of magnitude obtained by the analysis of the Cologne experiment. In a classical framework, the energy corresponding to the passage from an e.m. subminkowskian metric to the minkowskian one can be regarded as the attainment of the "border" of the photon, where "border" must be understood in the sense of the wavefront of the single e.m. wavepacket (or, better, as the boundary of the coherence region).

Let us now consider the range of energies for which the e.m. metric, describing the propagation of an evanescent wave in a waveguide, is subminkowskian. On the basis of the above discussion, it is now $E < E_{\text{coh}}$. Although such a region is expected to extend up to $E = 0$, a moment of thought suggests that actually there is a lower energy bound E_{min}, corresponding to the finite curvature radius of the Universe. Indeed, no e.m. field can have a spatial extension greater than the Universe itself. The minimal value one can provisionally get for E_{min} from the analysis of the Cologne experiment is $E_{\text{min}} \sim 10^{-26}\,\mu eV = 10^{-65}$ gr, that is roughly the experimental upper limit on the photon mass m_{0}. This is compatible with the so-called De Broglie mass $m^{*} = h/Rc$, if one takes for the Universe radius the value $R \sim 10^{10}$ light years, so that $m^{*} = 10^{-49}$ gr $= 10^{-10}\,\mu eV(m_{0} < m^{*})$. Conversely, if one assumes $E_{\text{min}} \simeq m^{*}$, it would be possible to deduce the value of R. Although no definite conclusion can be presently drawn about the actual value of E_{min}, let us stress that, if so, one would have in hand the noticeable deduction of the value of a cosmological constant — the curvature radius of the Universe — from a local experiment, just as it happened for the gravitational constant from the Cavendish experiment.

7.2. An Application: Unified View to Cologne and Florence Experiments

7.2.1. *Superluminal Propagation and the Friis Law*

One of the main problems for a theoretical treatment of the superluminal photon propagation is due to the fact that it was observed in different kinds of experiments, which are not easily comparable. It is so quite impossible to state if the results of different experiments are compatible with each other.

As an application of the DSR formalism in the case of the e.m. interaction, it will be shown that two of the first performed experiments, namely, the 1992 Cologne experiment (we discussed in the previous subsection), and the 1993 Florence experiment[36] on the microwave propagation in air between two horn antennas, do admit a common interpretation.[42] This allows one to set intriguing connections between these two (*a priori* different) classes of experiments.

Schematic views of the Cologne and Florence experimental devices are given in Figs. 6.1 and 7.4, respectively. The same symbol L is used to denote the length of the reduced-section waveguide in the former case and the distance between the two horn antennas in the latter.

The main tool to be exploited is the formula which gives the efficiency η of a transmitting device, *i.e.* the ratio between the received and the emitted power:

$$\eta = \frac{P_R}{P_E} = \frac{A_R A_E}{\lambda^2 L^2} \tag{7.10}$$

(also known as *Friis law*[43]). Here, $P_{R(E)}$ is the received (emitted) power, $A_{R(E)}$ the (effective) area of the surface $\Sigma_{R(E)}$ of the receiver (emitter),

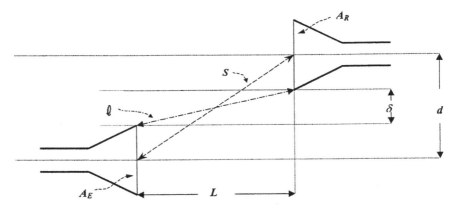

Fig. 7.4.   Schematic view of the two horn antennas used in the Florence experiment. L = distance between antennas; ℓ = distance between the upper border of the emitter and the lower border of the receiver; S = distance between the centers of the antenna surfaces; δ = normal distance between the receiver lower border and the emitter upper border; d = distance between the axes of antennas; A_E = area of the emitting antenna; A_R = area of the receiving antenna.

λ the wavelength of the transmitted electromagnetic signal, and L the distance between the surfaces Σ_R and Σ_E.

Some remarks are needed concerning the validity limits of the Friis law. It strictly holds true for large distance between emitter and receiver. However, it can be shown that it is valid approximately even for small distances (it can be derived indeed from the behaviour of the near field of the emitter). Moreover, Eq. (7.10) is usually interpreted as strictly holding for two faced antennas. But by its very derivation it holds too if the antennas are shifted (as in the Florence experiment), provided that the areas A_R, A_E are regarded as effective. In this case, their values are obtained by suitably projecting the real areas along the direction of propagation of the wave (*i.e.* of the Poynting vector).

In the Cologne experiment the rectangular waveguide with reduced section behaves like a high-pass filter with a cutoff frequency $\nu_c = 9.49\,\text{GHz}$. The experiment was carried out in undersized conditions, *i.e.* at an under-cutoff frequency $\nu = 8.70\,\text{GHz}$. In the narrow part of the waveguide (namely "under barrier" in the particle tunneling analogy), of length $L = 4\,\text{cm}$, the evanescent waves exhibit a decaying behaviour of the type

$$f \sim e^{-z/L_c} \tag{7.11}$$

where the z-axis is along the axis of the guide, and the cutoff length L_c (the penetration distance *of the wave*) is given by (cf. Eq. (7.7))

$$L_c = \frac{c}{2\pi}(\nu_c^2 - \nu^2)^{-1/2} = 1.259\,\text{cm}. \tag{7.12}$$

The propagation of such evanescent waves was found to occur at superluminal group velocity ($u = 1.03c$). The energy of the wave under the barrier decays according to the heuristic assumption

$$E = E_{\text{in}}e^{-z/L_0} \tag{7.13}$$

(cf. Eq. (7.5)), where E_{in} is the input energy, *i.e.* the energy entering the barrier, and L_0 is the penetration length *of the energy*. It follows obviously from Eq. (7.11)

$$L_0 = \frac{L_c}{2}. \tag{7.14}$$

Let us show that the application of the Friis law (7.10) to the Florence experiment permits to state that the two-horn antennas device, too, behaves like a high-pass filter. The frequency used in such an experiment was $\nu = 9.50\,\text{GHz}$. The area of the (rectangular) surface of either the emitting and the receiving antenna was $A = a \times h = 9\,\text{cm} \times 8\,\text{cm} = 72\,\text{cm}^2$ (with

a, h being the width and the height of antenna surface, respectively). If the antennas face each other, it is obviously $A = A_R = A_E$. Suppose now to shift the antennas orthogonally so that the distance between their axes is d. Let ℓ be the distance between the upper border of the emitter and the lower border of the receiver and S the distance between the centers of the antenna surfaces Σ (see Fig. 7.4) ($S = \sqrt{L^2 + d^2}$). In the following, we shall also need the normal distance between the receiver lower border and the emitter upper border $\delta = d - h$ (whence $\ell = \sqrt{L^2 + \delta^2}$).

The measurement runs were carried out for three different values of the distance L between the antennas and for different (integer) values of d. Superluminal propagation of the signal was observed only for the lowest value of L, $L = 21$ cm. In this case, the time taken by a light signal to travel such a distance is $t_0 = L/c = 0.7$ ns. For four values of d (namely $d = 12, 13, 14, 15$ cm) the measured time was $t = 0.6$ ns $< t_0$. Notice that such values of d correspond to a lack of space coherence between the two antennas (*i.e.* the projection of the emitter surface on the position of the receiver does not intersect the active surface of the receiver[a]). These values can be regarded as corresponding to an *inescapable superluminality*, namely as a superluminal condition wholly independent of the distance between antennas.

Let us estimate, by means of the Friis law, the cutoff frequency of the Florence experiment in such a condition. One gets from (7.10)

$$\nu = \eta^{1/2} \frac{L}{(A_R A_E)^{1/2}} c. \tag{7.15}$$

Applying Eq. (7.15) to the two-antenna system requires to introduce two corrective factors, related to the validity limits of the Friis law. First, since the antennas are shifted, the effective area A_R of the receiver is obtained by projecting the surface Σ_R along the direction of S, *i.e.* $A_R = A \cos \alpha = A(L/S)$.

Furthermore, the near-field effect must be explicitly taken into account. It results in an effective reduction of the emitter surface A_E, due to the reduced field amplitude at small distances. Such an effect was experimentally observed and measured for a horn antenna operating at $\nu \simeq 10$ GHz.[44] The near-field amplitude as a function of the displacement along the height

---

[a]Such a definition of space coherence is therefore a measure of the facing of the antenna surfaces, and is correlated to the geometric efficiency of an emitter-receiver system.

h of the antenna surface exhibits an almost Gaussian behaviour.[b] The field is significantly different from zero over a distance h_{eff} that is roughly half h: $h_{\text{eff}} \simeq h/2$. It follows that the effect of the near field is to reduce the area A of the emitter by the same factor, *i.e.* $A_E \simeq A/2$ must be put in the Friis law.

Equation (7.15) becomes therefore

$$\nu = \eta^{1/2} \frac{(2SL)^{1/2}}{A} c. \tag{7.16}$$

The cutoff frequency ν_c is obtained for $\eta = 1$:

$$\nu_c = \frac{(2SL)^{1/2}}{A} c. \tag{7.17}$$

The values obtained for ν_c range from $13.10\,\text{GHz}$ to $13.54\,\text{GHz}$. Since the operating frequency of the Florence experiment $\nu = 9.50\,\text{GHz}$ is under the cutoff frequency for every configuration of the apparatus, the conclusion is therefore that the whole system works as a high-pass filter in the evanescent mode, like the Cologne device.

Let us attempt to treat the Cologne experiment in terms of law (7.10), too. In this case, the waveguide with reduced section is to be considered as a system of two antennas with emitter surface Σ_E given by the larger section of the waveguide, and the reduced section as receiver surface Σ_R (large — narrow antenna system). Then, one has $A_E = h_E \times a_E = 2.296\,\text{cm} \times 1.016\,\text{cm} = 2.33\,\text{cm}^2$, $A_R = h_R \times a_R = 1.58\,\text{cm} \times 0.79\,\text{cm} = 1.25\,\text{cm}^2$ (where $h_i, a_i, i = E, R$, denote the height and the width of the waveguide). From Eq. (7.10), with $\eta = 1$, it is possible to evaluate the effective length L_0 of the two-antenna system corresponding to the undersized waveguide at the cutoff frequency $\nu_c = 9.49\,\text{GHz}$, by getting

$$L_0 = \frac{\nu_c}{c} \sqrt{A_R A_E} = 0.542\,\text{cm}. \tag{7.18}$$

Such a value is about half the cutoff length L_c:

$$L_0 \simeq \frac{1}{2} L_c \tag{7.19}$$

in agreement with Eq. (7.14).

---

[b] Actually, the field amplitude displays a structure (characteristic of the interference processes), but this is in influent to the present aims.

On the other hand, let us evaluate by the same approach the efficiency η of the undersized waveguide at the cutoff length L_c and at the operating frequency $\nu = 8.70\,\text{GHz}$. One gets from Eq. (7.10)

$$\eta_c = A_R A_E \left(\frac{\nu}{cL_c}\right)^2 = 0.156 \qquad (7.20)$$

which agrees with the value derived from the energy decay law (7.13) $(\eta_c = \exp(-L_c/L_0) = e^{-2} = 0.135)$.

7.2.2. *Two-Antenna System as a Barrier*

On the basis of the analogy between the two experiments, it is sensible to assume that (in the operating conditions of the Florence experiment) the two-antenna system behaves as a barrier. The role of L, namely, the length of the reduced portion of the waveguide, is played in this case by the minimal distance $\ell = \sqrt{L^2 + \delta^2}$ between the antennas. The length δ represents the space extension of the barrier. The minimal value of ℓ for which superluminal propagation is observed is $\ell = 21.4\,\text{cm}$, with $\delta = 4\,\text{cm}$.

From the previous arguments, such a barrier behaves as a high-pass filter at an operating under-cutoff frequency. Therefore, the energy can be assumed to exhibit a decaying behaviour similar to Eq. (7.13) for the undersized waveguide, namely

$$E = E_{\text{in}} \exp(-\ell/\ell_0). \qquad (7.21)$$

Here, E_{in} is the initial energy of the beam (*i.e.* the emitted energy), which for a monochromatic beam of $\mathcal{N}$ photons with frequency ν reads

$$E_{\text{in}} = \mathcal{N}h\nu \qquad (7.22)$$

and ℓ_0 is the critical length.

The formalism of permits to evaluate ℓ_0 for the Florence experiment, provided that one assumes — in analogy with the Cologne experiment — that, between the antennas, space-time is no longer Minkowskian but is endowed with an energy-dependent (spatially isotropic), deformed metric of the type (7.8).

Therefore, the maximal causal speed of light in any direction corresponding to (7.8) is given by (7.9), namely

$$u(E) = \frac{c}{b(E)} = c \left(\frac{E_{0,\text{e.m.}}}{E} \right)^{1/6}. \tag{7.23}$$

On the other hand, the (superluminal) speed u in this case is given by

$$u = \frac{\ell}{t} \tag{7.24}$$

where t is the measured time. By setting $\mathcal{N} = 1$ in Eq. (7.22), one gets therefore from Eqs. (7.21), (7.22), (7.23), (7.24) the following expression for ℓ_0:

$$\ell_0 = \frac{\ell}{\ln \left[\left(\frac{h\nu}{E_{0,\text{e.m.}}} \right) \left(\frac{ct}{\ell} \right)^6 \right]}. \tag{7.25}$$

This allows one to evaluate in an independent way, by a completely different approach, the cutoff frequency ν_c, which in this framework is given by (cf. Eq. (7.7))

$$\nu_c = \sqrt{ \left(\frac{c}{2\pi\ell_c} \right)^2 + \nu^2 } \tag{7.26}$$

where the cutoff length ℓ_c (related to the decay of the wave) is related to ℓ_0 by

$$\ell_c = 2\ell_0. \tag{7.27}$$

The values of ν_c vary in the range $9.515 \div 9.514\,\text{GHz}$. We recover then, by a completely different approach, the result that the two-horn antennas device behaves as a high-pass filter. The different value found, in this framework, for the cutoff frequency, is due to the fact that the present treatment is based on the spatially isotropic metric (7.8). The hypothesis of space isotropy, valid in the Cologne case, does no longer hold for the Florence experiment, due to the edge effects of the emitting antenna and the near-field behaviour. A more sound treatment requires to take into account both effects in the functional form of the space metric coefficients $b_i^2(E)$ ($i = 1, 2, 3$) (now to be assumed different).

It was therefore shown, by two different and independent approaches — one based on the Friis law and the other on the deformation of the spacetime — that the two-antenna system of the Florence experiment can be considered as a high-pass filter. The formalism of the deformed Minkowski space permits also to describe the behaviour of the Florence device as a barrier, with a decaying law for the energy of the evanescent-wave type, and therefore to interpret the experiment as a genuine tunneling one, in full analogy with the Cologne case. From the results of Chap. 6, one can state that (as in the Cologne experiment) such a wave (evanescent in the ordinary space-time) propagates at superluminal speed in the deformed Minkowski space with a *real* wavevector, *i.e.* it is an e.m. non-Lorentzian wave.

CHAPTER 8

ENERGY-DEPENDENT METRIC FOR GRAVITATION

8.1. Analysis of Experimental Data on Clock Rates

In order to describe the gravitational interaction in terms of an energy-dependent metric, the experimental input is provided by the measurements carried out by Alley[45] in 1975 and by Meystre and Scully[46] in 1983 on the relative rates of clocks in the gravitational field. In such experiments, the rate of cesium beam atomic clocks, raised to a higher gravitational potential by an aircraft, was compared (by short pulses of laser light) with the rate of similar clocks on the ground. The plane speed was such as to make as small as possible the special-relativistic effect on the measured time (*i.e.* it is possible to get small the terms in β^2 in such measurements).

It is worth stressing that the energy-dependent gravitational metric we are looking for is to be regarded as a *local* representation of gravitation, because the experiments considered took place in a neighborhood of Earth, and therefore at a small scale with respect to the usual ranges of gravity (although a large one with respect to the human scale).

Alley and co-workers verified to about 1.5% the validity of the time-dilation formula derived from the metric (3.4) for a moving clock, *i.e.*

$$\Delta \tau = \Delta t \sqrt{1 + \frac{2\phi}{c^2} - \frac{v^2}{c^2}}, \tag{8.1}$$

where v is the clock speed. They, therefore, utilized the data for a line measurement. On the contrary, they will here be used for a spectrum analysis, due to the fact that one is interested in deriving the expression of the time metric parameter b_0 as a function of the gravitational energy E.

The difference in elapsed proper time between the clock on the aircraft and the clock on the ground is

$$\Delta T = \sqrt{\eta_{00}} \, \Delta \tau - \Delta \tau = \Delta \tau (\sqrt{\eta_{00}} - 1) \tag{8.2}$$

or, on account of Eq. (2.2):

$$\Delta T = \Delta \tau (b_0 - 1). \tag{8.3}$$

By taking, for the test time parameter, the ansatz[a]

$$b_0(E) = \left[1 + \left(\frac{E}{E_0}\right)^n\right]$$
(8.4)

(with the factor 1 following from the requirement $\Delta T = 0$ for $E = 0$),
Eq. (8.3) becomes

$$\frac{\Delta T}{\Delta \tau} = \left(\frac{E}{E_0}\right)^n.$$
(8.5)

Moreover, it is reasonable to assume

$$E \equiv E(q)$$
(8.6)

where q is the aircraft quote over Earth surface in Alley's experiment. Since
it must be $E = 0$ for $q = 0$, it can be put

$$E(q) \propto q^k.$$
(8.7)

The simplest assumption is postulating for the energy the form

$$E = \varepsilon \left(\frac{g}{c^2}\right) q$$
(8.8)

where g is the gravity acceleration (for q in feet and ε in eV, $\frac{g}{c^2} = 3.324 \times 10^{-17} (\text{feet})^{-1}$) and ε is an energy characteristic of the system considered.
This agrees with the standard form of the energy of a body of mass m at
height q in the Earth gravitational field, $E = mgq$, if we identify ε with the
rest energy of the body, $\varepsilon = mc^2$.

The most natural choice for ε (which, in our case, represents a characteristic energy of the time measuring device) is the energy gap $h\nu$ of the Cs^{133}
hyperfine transition ($h = 4.136 \times 10^{-15}$ eV $\cdot$ sec; $\nu = 9.193 \times 10^9$ sec^{-1}),
because it provides the time standard (which triggers the time-interval measurements, amplified by the atomic clock device). So, Eq. (8.8) becomes

$$E = \frac{h\nu}{c^2} gq.$$
(8.9)

Let us notice that, in principle, both the frequency ν and the speed c
may *a priori* depend on the quote q of the flying clock. For the frequency,
the reason is very easily seen because, on account of Eq. (8.1), it is

$$\nu(q) = \nu_0 g_{00}^{-1/2} = \nu_0 b_0^{-1}.$$
(8.10)

However, such a dependence on the altitude does not affect time measurements provided that the atomic transition occurs at a fixed quote. This is

---

[a]On the basis of an assumed analogy between the strong and the gravitational metric:
see Chaps. 10, 11.

just the case of the Alley experiment, where the proper time measurements were taken with the aircraft flying at a given altitude.

The dependence of c on q is more subtle. It is due to the fact that, according to the formalism of DSR, c in Eqs. (8.8) and (8.9) is no longer the speed of light (which we recall to be only considered the maximal causal speed for electromagnetic interactions), but is to be regarded as the maximal causal speed u_{grav} of gravitation, and therefore expressed by:

$$c(q) = u_{k_{\text{grav}}} = \frac{b_0}{b_k}c. \tag{8.11}$$

However, in this case, too, the dependence of the speed on q can be neglected. This can be seen as follows. First, one can assume, on physical grounds (by analogy with the case of the electromagnetic and the weak metrics, which exhibit the same functional dependence on the energy: see Chap. 9) that the energy-dependent metric for gravitation behaves as the strong metric (see Chap. 10). This implies that we have $b_1 = \text{const}$, $b_2 = \text{const}$ $(b_1 \neq b_2)$, $b_3(E) = b_0(E)$, so that (putting the constants equal to 1, without any loss of generality)

$$u_{1_{\text{grav}}} = u_{2_{\text{grav}}} = cb_0; \quad u_{3_{\text{grav}}} = c. \tag{8.12}$$

In the hypothesis that the change of energy one is interested into is along the $k = 3$ direction, it is possible therefore to set

$$u_{3_{\text{grav}}} = c(q) = c. \tag{8.13}$$

Let us notice, however, that in both cases the q-dependence of the frequency and the light speed can actually be neglected. Indeed, the change in the expression (8.9) of the energy due to either effect is:

$$E = \frac{h\nu_0}{c^2}gq\left(1 + \frac{g}{c^2}q\right)^{-3/2} \simeq \frac{h\nu_0}{c^2}gq \tag{8.14}$$

(being $\frac{g}{c^2} = 3.324 \cdot 10^{-17}\text{feet}^{-1}$).

Therefore, we can, in full generality, assume for E the form (8.14) (with c the light speed in vacuum and $\nu = \nu_0$). Thus, Eq. (8.5) takes the form

$$\frac{\Delta T_{\text{grav}}}{\Delta\tau} = \left[h\nu_0\left(\frac{g}{c^2}\right)\left(\frac{1}{E_0}\right)q\right]^n \tag{8.15}$$

where $\Delta\tau = 1$ day $= 8.640 \cdot 10^{13}$ ns, and the suffix "grav" means that the time dilation is due only to gravitational effects.

Actually, the experimental data on ΔT do contain a term of kinetic origin ΔT_{kin}, which must be subtracted out according to the formula:

$$\Delta T_{grav} = \Delta T + \Delta T_{kin} \tag{8.16}$$

with[b]

$$\Delta T_{kin} = \left(\frac{v_0}{c}\right)^2 \frac{\Delta \tau}{2}. \tag{8.17}$$

With this proviso, we can rewrite Eq. (8.15) as

$$\frac{\Delta T_{grav}}{\Delta \tau} = A^n q^n \tag{8.18}$$

with

$$A = h\nu_0 \left(\frac{g}{c^2}\right)\left(\frac{1}{E_0}\right). \tag{8.19}$$

Putting $y = \frac{\Delta T_{grav}}{\Delta \tau}$, $x = q$, we get therefore the fit function

$$y = A^n x^n. \tag{8.20}$$

The values of the parameters n, A obtained by the fit are[47]

$$n = 0.9375 \pm 0.0047;$$
$$A^n = (54.000 \pm 1.149) \cdot 10^{-4} \, (\text{feet})^{-n}, \tag{8.21}$$

with $R^2 = 0.9519$.

8.2. Local Metric of Gravitation

Let us discuss the physical consequences of the above fit.

According to the results of the fit, it can be concluded that the energy-dependent time coefficient of the gravitational metric reads (cf. Eq. (8.4))

$$b_0(E) = 1 + \frac{E}{E_0} \tag{8.22}$$

---

[b]Let us notice that the kinetic correction does still have the standard expression, since the e.m. energy-dependent metric reduces to the Minkowskian one for the energy range of the Alley experiment, $E > E_{0em} = 5\,\mu eV$ (see Chap. 7).

where the gravitational energy threshold E_0 is given by

$$E_0 = E_{0,\text{grav}} = \frac{1}{A} h\nu_0 \left(\frac{g}{c^2}\right). \tag{8.23}$$

With the value of A obtained by the fit, one has

$$E_{0,\text{grav}} = 20.2 \pm 0.1 \,\mu\text{eV}. \tag{8.24}$$

Intriguingly enough, $E_{0,\text{grav}}$ is approximately of the same order of magnitude of the thermal energy corresponding to the $2.7°K$ cosmic background radiation in the Universe.

The time-dilation formula corresponding to the time coefficient (8.22) reads now

$$d\tau = \left(1 + \frac{E}{E_0}\right) dt \tag{8.25}$$

at variance with the Einsteinian one (derived from metric (3.4))

$$d\tau = \left(1 + \frac{E}{E_0}\right)^{1/2} dt. \tag{8.26}$$

As to the explicit form of the *spatial* part of the gravitational metric (on which the experimental data do not provide any information), two possibilities are open:

(1) *the spatial, 3-dimensional metric is Euclidean, i.e.* $b_1 = b_2 = b_3 = 1$; it is therefore

$$\eta_{\text{grav}}(E) = \text{diag}\left[\left(1 + \frac{E}{E_{0,\text{grav}}}\right)^2, -1, -1, -1\right]. \tag{8.27}$$

(2) *the spatial metric is anisotropic and energy-dependent, i.e.* the 4-dimensional metric has a structure similar to the strong one (see Chap. 10), namely

$$\eta_{\text{grav}}(E) = \text{diag}\left[\left(1 + \frac{E}{E_{0,\text{grav}}}\right)^2, -b_1^2, -b_2^2, -\left(1 + \frac{E}{E_{0,\text{grav}}}\right)^2\right] \tag{8.28}$$

with, in general, $b_1^2 \neq b_2^2$.

The gravitational metric is *overminkowskian*, since it attains the Minkowskian limit from above (for $E > E_{0,\text{grav}}$).

8.3. How Fast Do Travel Gravitational Effects?

The problem of the speed of transmission of gravitational effects is an old one. It can be traced back to Laplace, who, in his monumental work "*Mecanique Celeste*" in 1825, estimated $u_{\mathrm{grav}} \geq 10^8\,c$ from astronomical observations on the Moon-Earth-Sun system.

Let us clarify what we mean by "speed of gravitational effects". As is well known, General Relativity predicts the existence of gravitational waves, *i.e.* weak disturbances of the space-time metric[c] (obeying a Helmholtz wave equation) which propagate at the speed of light. Indirect confirmation of this speed was provided by Taylor and Hulse by the analysis of binary pulsars. Gravitational radiation does admit retarded-potential solutions of electromagnetic type. It therefore describes propagation of the perturbations of a *static* (or near static) gravitational potential field.

On the contrary, how much is the propagation speed *of the gravitational force*? By this we mean the speed at which *variations of the gravitational force* do travel. It answers the question of how much time a target body will take to respond to the acceleration of a source mass. Such a time is obviously zero, in Newtonian mechanics Borrowing a beautiful analogy from Van Flandern and Vigier,[48] let us consider a buoy floating on sea surface. The buoy is connected by a chain to an anchor holding it in place. If the anchor is moved, the chain causes the buoy to move too. In turn, the buoy motion sets off water waves. Translated in gravitational language, the anchor is the source mass, the chain is the gravitational force, the buoy the target mass. The water waves caused by the buoy motion (induced by the anchor motion) travel at the sound speed in water, and are the analogous of gravitational waves: there is no connection between their speed and the speed of transmission of the force field from the anchor to the buoy by the chain. Variations of the gravitational force (namely, variations of the whole space-time geometry) originate from acceleration of the source mass; gravitational waves (*i.e.* small ripples of the space-time geometry) originate from acceleration of the target body.

We want now discuss the problem of the speed of the gravitational force in the framework of DSR. According to the general formalism we developed in Chap. 4, the maximal gravitational causal speed depends on the metric coefficients. Then, if the gravitational metric is spatially Euclidean, *i.e.* it is given by Eq. (8.27), there is only one (maximal) speed of propagation of

---

[c]In fact, gravitational radiation is a fifth order effect in v/c.

gravitational signals, namely

$$u_{\text{grav}} = \left(1 + \frac{E}{E_{0,\text{grav}}}\right) c. \tag{8.29}$$

If the metric is spatially anisotropic (Eq. (8.28)), more u_k are possible. In particular, if $b_1^2 = b_2^2 = 1$, we have, according to Eqs. (2.9), (3.1) and (3.2),

$$u_{1_{\text{grav}}} = u_{2_{\text{grav}}} = c\left(1 + \frac{E}{E_{0,\text{grav}}}\right); \quad u_{3_{\text{grav}}} = c. \tag{8.30}$$

An estimate of the lower limit of the gravitational speed u_{grav} can be given by considering, for the energy E, the rest energy associated to the gravitational object of minimal mass constituting matter, *i.e.* the electron ($m_e \simeq 0.5\,\text{MeV}$). Replacing this mass value in Eq. (8.28), we get[47]

$$u_{\text{grav}} \geq \left(1 + \frac{m_e}{E_{0,\text{grav}}}\right) c = 2.5 \cdot 10^{10} c. \tag{8.31}$$

Although such a value for the speed of propagation of the gravitational signals seems paradoxical, it is actually in astonishing agreement with the results of the work by Van Flandern[49] about the acceleration of binary systems. The analysis of these data puts the lower limit $2 \cdot 10^{10}\,c$ on the speed of propagation of the gravitational force.[d] Such a result of a superluminal propagation of gravitational effects was also confirmed in 1997 by an experiment by Walker and Dual.[50]

8.4. Discussion of the Einstein versus Bohr Gedankenexperiment

As is well known, at the Sixth Solvay Conference in 1930 Einstein suggested a "Gedankenexperiment" in order to criticize the Heisenberg time-energy uncertainty principle.[51] The experiment involves a box containing photons, provided with a shutter controlled by a clock. The shutter is opened for a short time Δt, and some photons escape from the box. The variation in energy of the box, ΔE, can be therefore measured at any accuracy by weighting the box before and after the opening of the shutter.

Bohr[52] replied to Einstein's criticism by showing that the validity of the uncertainty principle is preserved in Einstein's experiment, provided that

---

[d]For reader's convenience, we recall that the equation for the gravitational speed used by Van Flandern is $v_g = \left[\frac{12\pi^2}{p\dot{p}}\left(\frac{a}{c}\right)\right] c$, where p is the period of the orbit of the binary pulsar, $\dot{p}$ is its time variation, and a is the major semiaxis of the orbit. The data refer to the binary pulsars PSR1913 + 16 and PSR1534 + 12.

one uses the appropriate time spread Δt which can be derived from g_{00} as given by the metric (3.4). His reasoning is as follows:

(1) Weighting the box requires the reading of a scale pointer with an accuracy Δx. Then, according to the position-momentum uncertainty principle, the box momentum acquires an uncertainty $\Delta p \gtrsim \hbar/\Delta x$.

(2) If the weighting takes a time T, and the mass change of the box is Δm, the corresponding momentum variation Δp_m (due to the mass variation) is $(g\Delta m)T$ (with g being the gravity acceleration) and is, in general, much larger than Δp:

$$(g\Delta m)T \gg \Delta p \rightarrow (g\Delta m)T \gg \hbar/\Delta x. \tag{8.32}$$

(3) General Relativity implies that a vertical position change Δx in a gravitational field causes a change ΔT in the clock rate, given by

$$\Delta T = \frac{gT\Delta x}{c^2} \tag{8.33}$$

whence

$$T = \frac{\Delta T}{g\Delta x}c^2. \tag{8.34}$$

Replacing the above relation in Eq. (8.32) yields

$$(\Delta m)c^2\Delta T \gg \hbar \tag{8.35}$$

namely the time-energy uncertainty relation (since $(\Delta m)c^2 = \Delta E$).

Notice that Eq. (8.33) is obtained from metric (3.4) by neglecting terms of the order c^{-4}, because

$$\frac{\Delta T}{T} = \sqrt{1 + \frac{2\phi}{c^2}} - 1 \simeq \frac{\phi}{c^2} = \frac{g\Delta x}{c^2}. \tag{8.36}$$

The same argument of Bohr can be applied here to the gravitational metric (8.27), or (8.28). In this case, with (8.22) for $b_{0,\text{grav}}(E)$, one gets:

$$\frac{\Delta T}{T} = b_{0,\text{grav}}(E) - 1 = \frac{E}{E_{0,\text{grav}}}. \tag{8.37}$$

Then, we have

$$\Delta p_m \Delta x = (g\Delta m)\Delta T\Delta x\frac{E_{0,\text{grav}}}{E} \tag{8.38}$$

and (on account of the fact that $E > E_{0,\text{grav}}$) it cannot be stated that, in general such a quantity is greater than (or at least of the order of) $\hbar$.

We can therefore conclude that, in the DSR framework, Bohr's rebuttal of the Einstein argument invalidating the time-energy uncertainty principle is no longer valid. This example shows that DSR may also affect quantum mechanics.

8.5. Einstein and the Energy-Dependence of the Light Speed

At this point it seems worth to recall the historical trek that led Einstein to formulating his equations which are at the very basis of General Relativity.[53] The starting point in 1911 was the well-known formula for the variation in frequency of a light signal in presence of a gravitational field (*i.e.* the gravitational redshift):

$$\nu_1 = \nu_2 \left(1 + \frac{\phi}{c^2}\right), \quad \phi = \phi_1 - \phi_2 \tag{8.39}$$

where ν_i, ϕ_i $(i = 1, 2)$ are, respectively, the local frequency and gravitational potential at point P_i. From such equation, Einstein concluded that the speed of light, too, must vary in the same way:

$$c_1 = c_2 \left(1 + \frac{\phi}{c^2}\right) \tag{8.40}$$

where the speed variation Δc is assumed to be small $(\Delta c/c \ll 1)$, so that c^2 is still a true invariant (see Sec. 1.1).

Then, in 1912 Einstein faced the task of finding the equation expressing the variation of c determined by a static matter distribution ρ, and assumed the following form

$$\Delta c = kc\rho \tag{8.41}$$

where Δ is the (3-dimensional) D'Alembert operator and k a constant.

Later, he realized that the source term in Eq. (8.41) is in general the energy density of the system (including therefore also the energy density of an electromagnetic field, and that of the gravitational field itself — whence the need of a nonlinear generalization of Eq. (8.41)).

However, Einstein was worried by the problem of giving up the principle of the constancy of the light speed. As a matter of fact, this was one of the points used by M. Abraham to question the theory of relativity. Late in 1912 G. Nordström suggested to abandon the idea of a dependence of c on the gravitational potential ϕ, and to assume instead the dependence of mass on ϕ. In 1913 Einstein and A.D. Fokker concluded that the Nordström theory of gravitation is a special case of Einstein–Grossmann's theory, constrained by the requirement of the constancy of the speed of light. By the way, in the static limit the gravitational field equations reduce to

$$\Delta\phi = -\kappa\rho \qquad (8.42)$$

which is well known to yield the Newtonian limit of Einstein's field equations.

Let us stress that, for the gravitational metric (8.27) corresponding to the Newtonian case, the gravitational maximal causal speed u_{newt} is just

$$u_{\text{newt}} = cb_0 = c\sqrt{1 + \frac{2\phi}{c^2}} \qquad (8.43)$$

which is essentially Eq. (8.29). In this case, Einstein finds the right answer, by asking the wrong question. We can state that the formalism of DSR amounts to follow Einstein's legacy before the Nordström proposal: *the speed of an interaction* (in this case, the gravitational one) *may well depend on the energy.*

We recall that a possible energy dependence of the speed of light has been considered both from a theoretical and an experimental side. Indeed, it has been shown[54] that quantum gravity may lead to modifications of the dispersion relation for high-energy photons, introducing a dependence on the energy. Such an effect may be observed in pulsar radiation.[55] Furthermore, astronomical observations on the absorption of quasar radiation from the interstellar gas suggest to revive the early Dirac hypothesis[56] of a time variation of the fine-structure constant — in particular of the light speed.[57] In turn, such a variation might be due to changes in the energy density of the vacuum during the Universe expansion.[58] Multidimensional theories like the superstring and the Kaluza–Klein ones do predict an energy dependence of the fundamental constants.[59,60] It would be therefore interesting to compare the DSR predictions with the above theoretical speculations and experimental observations, in particular with the results of quantum gravity and string models.

CHAPTER 9

WEAK INTERACTION

Let us now discuss the leptonic interaction. In order to derive the phenomenological weak metric, a purely leptonic decay has been analyzed, *i.e.* that of the meson K_s^0, whose meanlife is experimentally known in a wide energy range $(30 \div 350 \, \text{GeV})$.[61,62] This is an almost unique case, motivated by the fact that the K_s^0 decays violate charge conjugation-parity (CP) invariance, and therefore (by virtue of the celebrated CPT theorem) do provide indirect evidence for violation of time reversal.

Moreover, it is still assumed — as in the electromagnetic case — a spatially isotropic metric. As it will be seen in the following, the isochronism with the usual Minkowski metric (*i.e.* $b_0^2 = 1$) is derived from the phenomenological analysis, without any need of postulating it.

In this case, one has to exploit the law of time dilation in the form reported in Table 5.2 of Chap. 5, written here again for reader's convenience:

$$\Delta t = \frac{\Delta t_0}{\left[1 - \left(\frac{b}{b_0} \right)^2 + \left(\frac{b}{b_0} \right)^2 \left(\frac{m_0}{E} \right)^2 \right]^{1/2}}. \tag{9.1}$$

Although the known data on K_s^0 lifetime belong to two different measurements, taken in two different experiments, they had been made statistically homogeneous, thus allowing one to treat them as a single set of data in the analysis by relation (9.1).[63]

The main steps in deriving the weak metric are as follows[4,63]. Firstly, the data interpolation via the minkowskian linear law (see Table 5.2 in Chap. 5) is compared with the interpolation via Eq. (9.1), thus providing the conclusion that the minkowskian law does not yield results consistent with a constant metric. Then, one carries out — still via Eq. (9.1) — an interpolation of the experimental data in subintervals at different energies, so finding the values of b^2 and b_0^2 at various energies. Such a procedure allows one to conclude that b_0^2 does not vary with energy — unlike b^2 — at least in the range considered. The interpolation of the values of parameters b^2 and b_0^2 permits then to derive an explicit expression of the metric, which yields the phenomenological description of leptonic interaction, regarded as an interaction not derivable from a potential.

The results of the fit yield the following metric[4]:

$$\eta_{\text{weak}}(E) = \text{diag}\,(b_{0,\text{weak}}^2(E), -b_{\text{weak}}^2(E), -b_{\text{weak}}^2(E), -b_{\text{weak}}^2(E));$$
$$b_{0,\text{weak}}^2(E) = 1;$$
$$b_{\text{weak}}^2(E) = \begin{cases} (E/E_{0,\text{weak}})^{1/3}, & 0 \leq E \leq E_{0,\text{weak}} = 80.4 \pm 0.2\,\text{GeV} \\ 1, & E_{0,\text{weak}} < E \end{cases}.$$

$$(9.2)$$

The meaning of $E_{0,\text{weak}}$ is still the energy value at which the metric becomes Minkowskian. Although it is impossible to ascertain with certainty, in this framework, if the metric, once become Minkowskian at $E_{0,\text{weak}}$, does last so, it is worth stressing that such a value of the energy corresponds to the mass of the W-boson, through which the K_s^0-decay does occur. Moreover, $\eta_{\text{weak}}(E)$ is subminkowskian, like the electromagnetic metric.

We therefore recover the well-known fact that, within the unified model of electromagnetic and weak interactions by Glashow–Weinberg–Salam (based, as is well known, on the SU(2) × U(1) symmetry), $E_{0,\text{weak}}$ is the energy scale at which the weak and electromagnetic interactions are mixed.

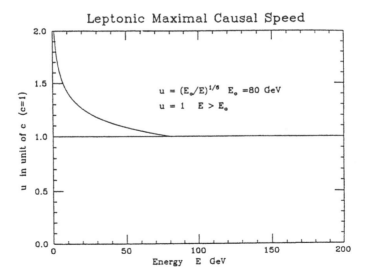

Fig. 9.1.   Plot of the leptonic maximal causal speed vs. energy (in units of c).

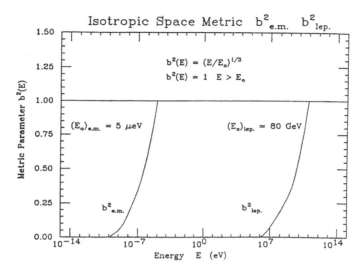

Fig. 9.2. Showing the behaviour of the electromagnetic and leptonic spatial parameters, in the hypothesis of a spatially isotropic metric.

The new interpretation provided by the DSR formalism is that $E_{0,\text{weak}}$ is the energy scale at which the weak and electromagnetic metrics measure space-time separations in the same way.

Therefore, it is not surprising that, using the same interpolation conditions, the leptonic metric (9.2) has the same form of the e.m. metric (7.8). However, we do not want to further pursue here the possible implications of this phenomenological analysis, which will be considered in Chap. 11.

From the metric (9.2) of the weak interaction, it is possible — as already done for the e.m. case — to get the leptonic maximal causal speed u_{lep}, on account of Eq. (3.8).

The expression of $u_{\text{lep}}(E)$, in units of c, is given by:

$$u_{\text{lep}}(E) = \begin{cases} (E/E_{0,\text{weak}})^{1/6}, & 0 \leq E \leq E_{0,\text{weak}} = 80.4 \pm 0.2\,\text{GeV} \\ 1, & E_{0,\text{weak}} < E \end{cases} \quad . \quad (9.3)$$

The behaviour of the leptonic maximal speed (9.3) is plotted in Fig. 9.1.

Finally, Fig. 9.2 shows (in an energy logarithmic scale) the spatial parameters of the isotropic electromagnetic metric (7.8) and of the isotropic leptonic metric (9.2). Notice that, due to the energy scale used, it cannot be stated that either parameter vanishes at zero energy (at least at the present status of development of the DSR formalism).

CHAPTER 10

STRONG INTERACTION

In order to discuss, in the DSR framework, the case of the hadronic inter-action, we shall consider the phenomenon of Bose–Einstein (BE) corre-lation for bosons (in practice, mesons), produced by such interaction in high-energy collisions.[64] The BE effect (well-known to be independent from the nature of the colliding particles) allows one to start, *a priori*, from an anisotropic metric, and so to check directly from the analysis of the phe-nomenon whether there is isotropy or not.

10.1. The Bose–Einstein Correlation

The so-called Bose–Einstein correlation in pion production in high-energy collisions is the phenomenon wherein pairs of identical bosons show a higher probability of emission at small opening angles — or, equivalently, at small relative momenta — than pairs of non-identical particles. Its name is due to its early interpretation (for equally charged pions) as a manifestation of their BE statistical properties, namely as a second-order interference effect of the wave functions of the particles, with the consequent requirement of a total wave function symmetric under particle exchange.

The relevant quantity is the BE second-order correlation function $C_{(2)}$, defined by

$$C_{(2)} = \frac{P(p_1, p_2)}{P_0(p_1, p_2)} \qquad (10.1)$$

where p_i $(i = 1, 2)$ is the four-momentum of the ith-particle, $P(p_1, p_2)$ is the two-particle probability density subjected to BE symmetrization and $P_0(p_1, p_2)$ is the corresponding quantity in absence of correlation. In the "canonical" treatment of the BE effect, the probability $P(p_1, p_2)$ is given by

$$P(p_1, p_2) = \int |\Psi_{12}^{BE}(x_1, x_2; r_1, r_2)|^2 \rho(r_1)\rho(r_2) d^4 r_1 d^4 r_2 \qquad (10.2)$$

where Ψ_{12}^{BE} is the amplitude for a boson pair to be produced at r_1 and r_2 and detected at x_1 and x_2, and $\rho(r_i)$, the space-time distribution of the ith-boson source, is assumed to have the Gaussian form

$$\rho(r_i) = \frac{1}{4\pi^2 R^4} \exp\left(-\frac{r_i^2}{2R^2}\right). \qquad (10.3)$$

The result is the following expression of $C_{(2)}$

$$C_{(2)} = 1 + \exp(-Q^2 R^2). \qquad (10.4)$$

We adopted the Goldhaber convention by introducing the four-momentum transfer

$$\begin{aligned} Q &\equiv (p_1 - p_2) = (q_0, \mathbf{q}); \\ Q^2 &= q_0^2 - q_t^2 - q_l^2, \end{aligned} \qquad (10.5)$$

where p_1, p_2 are the momentum fourvectors of the two bosons, and q_t and q_l are, respectively, the components of $\mathbf{q}$ orthogonal and parallel to the vector $\mathbf{n} = (\mathbf{P}_1 + \mathbf{P}_2)/|\mathbf{P}_1 + \mathbf{P}_2|$. From kinematical considerations it follows that $q_\ell \simeq q_o$, and therefore the experimental correlation function depends only on q_t. However, formula (10.4) is unable to account for the experimental findings, and, in order to get a satisfactory agreement with the experimental data, one is forced to introduce an *"ad hoc"* *"incoherence (or chaoticity) parameter"* λ ($0 \leq \lambda \leq 1$), physically interpreted as the fraction of pairs of identical particles that appear to interfere (and are therefore correlated):

$$C_{(2)} = 1 + \lambda \exp(-Q^2 R^2). \qquad (10.6)$$

10.2. DSR Treatment of BE Correlation

The presence of the parameter λ (totally alien to the model) is one of the drawbacks of the "canonical" treatment of the BE correlation. Moreover, there is evidence — first provided by Gaspero[65,66] — that the experimental value of the Bose–Einstein correlation function in some cases is actually greater than 3.[67] The value of the chaoticity λ obtained in correspondence is $\lambda \gg 1$, thus invalidating the standard meaning (*i.e.* fraction of pairs of bosons which are correlated) attributed to such a parameter in the usual treatment of the BE phenomenon. Both these points can be successfully dealt with in the DSR description of the BE correlation.[68,69]

Let us briefly review the main steps in the derivation of the "deformed" correlation function.

The basic assumption is that the strong interactions responsible of the BE phenomenon may be nonpotential and/or nonlocal, and therefore (according to the DSR formalism) space-time is deformed inside the interaction region, where pions are produced (the *"fireball"* of the BE correlation). The spatial parameters b_k of the deformed metric η describe the spatial deformation of the fireball, whereas the time parameter b_0 is related to its lifetime. Precisely, the physical parameters a_μ of the fireball are related to the b_μ's by

$$a_k = \hbar c b_k; \quad a_0 = \hbar b_0. \tag{10.7}$$

The deformation of the Minkowski metric induces a change in the phase factors of the two-boson symmetrized wave function (essentially due to the deformed scalar product $*$), which therefore becomes

$$\tilde{\Psi}_{12}^{BE}(x_1, x_2; r_1, r_2)$$
$$= \frac{1}{\sqrt{2}}\left\{e^{ip_1*(x_1-r_1)}e^{ip_2*(x_2-r_2)} + e^{ip_1*(x_1-r_2)}e^{ip_2*(x_2-r_1)}\right\}. \tag{10.8}$$

Moreover, in order to account for a possible anisotropic distribution of the boson subsources inside the total source, one has to consider a four-vector source function, defined as

$$\tilde{\rho}_\mu(r) = \frac{1}{4}a_\mu^4 \exp\left(-\frac{r^2 a_\mu^2}{2}\right). \tag{10.9}$$

As a consequence, the correlation probability, too, is different for different space-time directions, and reads (ESC off)

$$\tilde{P}(p_1, p_2)(\mu) = \int |\tilde{\Psi}_{12}^{BE}(x_1, x_2; r_1, r_2)|^2 \tilde{\rho}_\mu(r_1)\tilde{\rho}_\mu(r_2) d^4 r_1 d^4 r_2. \tag{10.10}$$

Here, obviously, the momenta p_i are those measured in the laboratory frame, *i.e.* in full Minkowskian conditions. The information on the deformation inside the fireball is entirely contained in the deformation of the scalar product. Such an effect, whereby particles do keep memory of the anisotropy of the forces which produced them, can be regarded as a kind of *hadronic Einstein–Podolski–Rosen effect*.

Replacing Eqs. (10.8), (10.9) in Eq. (10.10), one gets, by calculations similar to the standard ones, the following expression for the deformed second-order BE correlation function, $\tilde{C}_{(2)}$:

$$\tilde{C}_{(2)} = 1 + |\tilde{F}_\mu|^2 \tag{10.11}$$

namely, a different correlation function for each space-time direction. According to (10.10), $\tilde{F}_\mu$ is essentially the generalized Fourier transform of the anisotropic distribution function of the sources inside the interaction region.

Since, in the experiments, one measures a *global* BE correlation function, in order to compare the theoretical predictions of the DSR anisotropic model with the experimental data, Eq. (10.11) must be suitably averaged on all space-time directions. This can be done by assuming that the squared norm in (10.11) is the (absolute value of) the norm of a four vector. Such a norm can be considered either in the standard or in the deformed Minkowski space. In the former case, one gets

$$\tilde{C}_{(2)}^{ISO} = 1 + |\tilde{F}_\mu^* g^{\mu\nu} F_\nu|$$

$$= 1 + \left| \exp\left(-\frac{Q^{\tilde{2}}}{a_0^2} \right) - \sum_{k=1}^{3} \exp\left(-\frac{Q^{\tilde{2}}}{a_k^2} \right) \right| \qquad (10.12)$$

where $Q^{\tilde{2}}$ is the deformed norm of the momentum transfer

$$Q^{\tilde{2}} = Q_\mu \eta^{\mu\nu} Q_\nu = b_0^2 Q_0^2 - b_1^2 Q_1^2 - b_2^2 Q_2^2 - b_3^2 Q_3^2. \qquad (10.13)$$

The upper index "ISO" in (10.12) means that the average $\tilde{C}_{(2)}^{ISO}$ is an *isotropic* one: it has been performed *outside* the interaction region, where the space-time has its usual Minkowskian structure. The particles involved in the BE process do keep memory of the nonlocal, anisotropic forces that produced them through the different form of the correlation function, and the presence of the deformed metric parameters in the exponential function.

The other possibility is to average the deformed correlation function (10.11) *inside* the interaction region, *i.e.* by using the deformed metric η. One gets

$$\tilde{C}_{(2)}^{AN} = 1 + |\tilde{F}_\mu^* \eta^{\mu\nu} F_\nu|$$

$$= 1 + \left| b_0^2 \exp\left(-\frac{Q^{\tilde{2}}}{a_0^2} \right) - \sum_{k=1}^{3} b_k^2 \exp\left(-\frac{Q^{\tilde{2}}}{a_k^2} \right) \right| \qquad (10.14)$$

where "AN" now means "anisotropic". The average (10.14) corresponds to differently weighting each direction, according to the fact that the deformed metric (3.3) implies a "renormalization" of the lengths, different for each direction.

The two averaged correlation functions predict different limits for the peak values. Indeed, in the case of $\tilde{C}_{(2)}^{AN}$ (10.14), it is easy to see that, in the Minkowskian limit (since every b_μ^2 is of the order of $1/3$),[a] the upper limit of $\tilde{C}_{(2)}^{AN}$ ($Q = 0$) is

$$\tilde{C}_{(2)}^{AN,\max} = 1 + \frac{1}{3} + \frac{1}{3} + \frac{1}{3} - \frac{1}{3} = 1.67. \tag{10.15}$$

In the isotropic case, one gets instead

$$\tilde{C}_{(2)}^{ISO,\max} = 1 + 1 + 1 + 1 - 1 = 3. \tag{10.16}$$

In the DSR framework, therefore, the peak value $h_{(2)}^{\text{peak}}$ of the correlation function is expected to vary within the values

$$1.67 \leq h_{(2)}^{\text{peak}} \leq 3. \tag{10.17}$$

10.3. Deformed BE Correlation Function and the Strong Metric

In order to derive the expression of the strong metric, one has to consider the anisotropic correlation function (10.14) (*i.e.* to take the average inside the interaction region). On account of Eq. (10.5), $\tilde{C}_{(2)}^{AN}$ can be expressed in terms of the transverse momentum transfer q_t as

$$\tilde{C}_{(2)}^{AN} = 1 + b_0^2 \left[e^{-q_t^2} - b_1'^2 e^{-q_t^2/b_1'^2} - b_2'^2 e^{-q_t^2/b_2'^2} - b_3'^2 e^{-q_t^2/b_3'^2} \right]$$
$$b_k' = \frac{b_k}{b_0}, \quad k = 1, 2, 3. \tag{10.18}$$

The products have been carried out classically in the usual Minkowski space (since the definition of q_t is the experimental one).

Equation (10.18) is then used to interpolating the data concerning the mesons produced in proton-antiproton annihilation at total energy given by the measurements taken in the UA1 "ramping run" in 1984.[70] The annihilation has been chosen in order to have the same energy in the center-of-mass and in the laboratory frames, so to refer the values of the parameters b_μ^2 to a unique energy value.

The results thus obtained for the b_μ^2's from Eq. (10.18) show that the phenomenon of BE correlation is indeed spatially anisotropic, so that the related metric is anisotropic. Therefore, we can state that spatial anisotropy

---

[a]This is due to the fact that, before the deformation, the spatial shape of the fireball can be considered spherical, with unit radius, so that the mean squared values of the spatial parameters b_k^2 (which are related to the spatial sizes of the interaction region) are of the order $1/3$. Moreover, since both the spatial deformation of the source and the appearance of the time parameter b_0^2 are to be ascribed to the same effects, it is expected that b_0^2, too, is of the same order of magnitude of the spatial parameters.

is, in general, a peculiar feature of hadronic interactions. In spite of the phenomenological nature of the correlation function (10.18), it allows one to get new (and even unforeseen) results, like the loss of axial symmetry of the BE phenomenon (commonly assumed from the very beginning in its standard treatment), the geometrical description, via the parameters b_μ, of the annihilation region, and the q_t value at which the correlation effect begins.[68] What's more, this analysis yields a strong evidence that hadronic interaction can be treated neither in terms of a metric with constant parameters, nor even in terms of a (standard) Minkowski metric.

By interpolating the experimental data, there follows that the explicit form of the metric η which describes the hadronic interaction, as a function of $E = \sqrt{s}$ (see Eq. (3.5), Sec. 3.2), reads

$$\eta_{\text{strong}}(E) = \text{diag}(b_{0,\text{strong}}^2(E), -b_{1,\text{strong}}^2, -b_{2,\text{strong}}^2, -b_{3,\text{strong}}^2(E))$$
$$b_{0,\text{strong}}^2(E) = b_{3,\text{strong}}^2(E)$$
$$= \begin{cases} 1, & 0 \le E \le E_{0,\text{strong}} = 367.5 \pm 0.4\,\text{GeV} \\ (E/E_{0,\text{strong}})^2 & E_{0,\text{strong}} < E \end{cases}$$
$$b_{1,\text{strong}}^2 = (\sqrt{2}/5)^2$$
$$b_{2,\text{strong}}^2 = (2/5)^2. \tag{10.19}$$

In this case, too, $E_{0,\text{strong}}$ is to be meant as the energy value at which the metric becomes Minkowskian.

Moreover, for $E < E_{0,\text{strong}}$, the hadronic parameters $b_{0,\text{strong}}^2$ and $b_{3,\text{strong}}^2$ take the Minkowskian value, which, for such energy range ($80.4\,\text{GeV} < E < 367.5\,\text{GeV}$), coincides with that of both e.m. and leptonic interactions.

Figure 10.1 shows the behaviour of the ratio $b_{0,\text{strong}}^2(E)/b_{3,\text{strong}}^2(E)$, whence it is easily seen that we can put $b_3^2(E) = b_0^2(E)$ (see Eq. (10.19)). In Fig. 10.2 all hadronic metric parameters b_μ are plotted together vs. energy, in order to emphasize the spatial anisotropy of the metric. Two spatial parameters are constant, but different in value, whereas the third one varies and is overminkowskian as energy increases.

Let us recall that the metric parameters are related to the space sizes and to the meanlife of the BE fireball. In particular, by using the minimum bias of UA1 with energy $\sqrt{s} = 630\,\text{GeV}$, one gets for a_0 the value

$$a_0 = \tau = (1.09 \pm 0.01) \times 10^{-24} s. \tag{10.20}$$

The corresponding "width" Γ of the fireball, given by

$$\Gamma \equiv \frac{\hbar}{a_0} = 0.61 \pm 0.02\,\text{GeV} \tag{10.21}$$

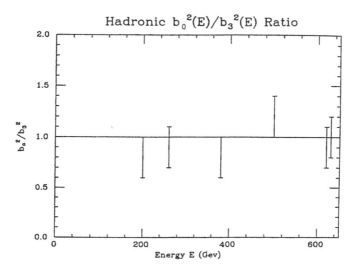

Fig. 10.1. Ratio of the hadronic metric parameters b_0^2/b_3^2.

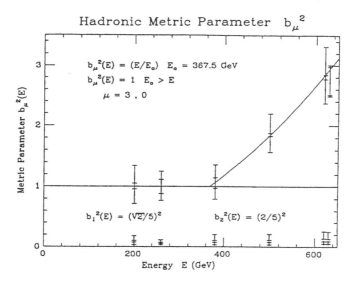

Fig. 10.2. Behaviour with energy of all four hadronic metric parameters b_μ^2.

is consistent with the experimental data, if one interprets Γ as that value of q_t such that $C_{(2)}^{\exp} \gtrless 1$ if $q_t \lessgtr \Gamma$.

Using the strong metric (10.19) it is possible to find the hadronic maximal causal speeds. Let us stress that the hadronic interaction provides an example of an anisotropic metric, and, therefore, there will be different maximal causal speeds $u_{k,\text{strong}}(E)$ corresponding to the different spatial directions.

They explicitly read, in units of c:

$$u_{1,\text{strong}}(E) = b_{0,\text{strong}}/b_{1,\text{strong}}$$
$$= \begin{cases} (5/\sqrt{2}), & 0 \leq E \leq E_{0,\text{strong}} \\ (5/\sqrt{2})(E/E_{0,\text{strong}}), & E_{0,\text{strong}} < E \end{cases} ;$$
$$u_{2,\text{strong}}(E) = b_{0,\text{strong}}/b_{2,\text{strong}} \qquad\qquad (10.22)$$
$$= \begin{cases} (5/2), & 0 \leq E \leq E_{0,\text{strong}} \\ (5/2)(E/E_{0,\text{strong}}), & E_{0,\text{strong}} < E \end{cases} ;$$
$$u_{3,\text{strong}}(E) = b_{0,\text{strong}}/b_{3,\text{strong}} = 1.$$

The fact that $u_{3,\text{strong}} = 1$, *i.e.* c, is already present in Fig. 10.1.

Lastly, Fig. 10.3 shows, for a comparison, the spatial parameters $b_{3,\text{weak}}^2$ (isotropic case) and $b_{3,\text{strong}}^2$ of the leptonic metric (9.2) and of the hadronic one (10.19), respectively. It is seen that the leptonic parameter exhibits a subminkowskian behaviour, unlike the hadronic one, which is overminkowskian. Moreover, both parameters take unity value in the range $80.4 \div 367.5\,\text{GeV}$, like in the e.m. case. Quite likely, the fact that — in the

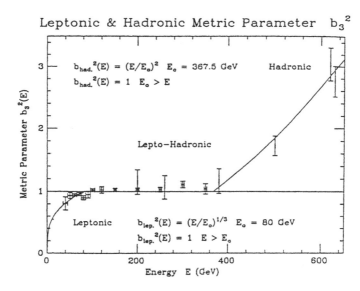

Fig. 10.3.   Comparing the spatial parameters b_3^2 for leptonic and hadronic metrics.

above energy range — the metric looks minkowskian for electromagnetic, weak and strong interactions, may explain why perturbation methods are so successful in describing both leptonic and hadronic reactions for such intervals of energy of the reaction final products (even when such methods would be *a priori* expected to fail, like in perturbative quantum chromodynamics).

Let us explicitly notice that the strong metric (10.19) is not always isochronous with the usual Minkowski metric ($b_0^2 = 1$) (which, as by now familiar, characterizes the electromagnetic interaction). Actually, it follows from (10.19) that it is $b_{0,\text{strong}}^2 \neq 1$ for $E_{0,\text{strong}} < E$.

Such a case is not new; indeed, as is well known, the same happens for the gravitational interaction, as shown e.g. by the various measurements of red or blue shifts of electromagnetic radiation in a gravitational field, or by the relative delays of atomic clocks put at different heights in presence of gravity.

As a final remark, it is worth noticing that — like a physical meaning can be attributed to $E_{0,\text{weak}}$ as the energy scale of the intermediate vector bosons for electroweak interactions — an analogous interpretation can be given to $E_{0,\text{strong}}$. A possible suggestion is that the value of $E_{0,\text{strong}}$ does represent the energy scale corresponding to the upper limit of the mass of the Higgs boson, that — as is well known — breaks the gauge invariance of the lepto-strong mixed interactions by endowing the weak interaction carriers with mass. On the other hand, in the DSR context, it is possible to give inertial mass to (at least) stable particles on the basis of Lorentz invariance breakdown (see Chap. 14). Therefore, such an interpretation of $E_{0,\text{strong}}$ requires a deep reflection on the possible links between gauge invariance and Lorentz invariance breakdowns, which, in our opinion, can be only dealt with in the framework of a five-dimensional formulation of DSR (see Chap. 15).

10.4. Hadronic Time Deformation

Let us investigate the possible implications of such an anisochronism of the hadronic metric. We denote by dt_{had} the time interval taken by a certain hadronic process for a particle at rest (*"hadronic clock"*). The same process, when referred to a Minkowskian electromagnetic metric, will take a time $dt_{\text{e.m.}}$ to happen. Borrowing methods and notation from the general theory of relativity, we can state that, for a particle at rest:

$$\frac{dt_{\text{had}}}{dt_{\text{e.m.}}} = \frac{1}{\sqrt{\eta_{00,\text{strong}}}}. \tag{10.23}$$

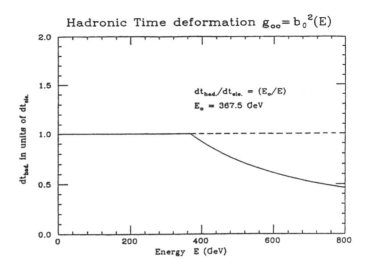

Fig. 10.4. Time deformation law in a hadronic field vs. the energy.

Since for the deformed metric (10.19) it is $\eta_{00,\text{strong}} = b^2_{0,\text{strong}}(E)$, we get

$$\frac{dt_{\text{had}}}{dt_{\text{e.m.}}} = \begin{cases} 1, & 0 \le E \le E_{0,\text{strong}} = 367.5 \pm 0.4\,\text{GeV} \\ E_{0,\text{strong}}/E, & E_{0,\text{strong}} < E \end{cases} . \qquad (10.24)$$

Equation (10.24) yields *the law of time dilation in a hadronic field*. Figure 10.4 shows the behaviour of law (10.24), *i.e.* the plot of dt_{had} vs. energy in units of $dt_{\text{e.m.}}$. It is easily seen that there is isochronism at low energies (*i.e.* physical processes have the same rate either when referred to a hadronic metric or to an electromagnetic one), whereas there is a time contraction at high energies. In other words, hadronic processes are faster when observed with respect to an electromagnetic metric.

10.5. A Geometric View to Confinement and Asymptotic Freedom

The hadronic time deformation law (10.24) provides us with an interesting representation of two fundamental features of strong interactions, *i.e.* asymptotic freedom and confinement of the hadronic constituents (say, quarks).

In deep inelastic scattering, where such properties of strong interaction are observed, the probe particles interact electromagnetically with the hadron constituents. When low-energy probe particles are involved — *i.e.* energy exchange occurs between the probe leptons (which undergo scattering) and the hadronic constituents (scattering centers) at energy values low with respect to the energy scale of strong interaction —, hadronic constituents behave essentially as free particles, that is they are "asymptotically free". On the basis of Eq. (10.24), such a fact can be interpreted in terms of equal time intervals for either the electromagnetic and the strong interaction, during which the same amount of energy is exchanged. Otherwise speaking, both e.m. and strong processes require the same time interval in order that particles exchange the same amount of energy. Therefore, in exchanging energy at such "low" values, hadron constituents behave exactly as the electromagnetic probes which, when scattered, do not keep any memory of the "bond" due to the strong interaction.

On the contrary, with increasing energy exchange between electromagnetic probes and quarks, Eq. (10.24) shows that *different* time intervals are needed to the electromagnetic interaction and the strong one in order to transfer the *same* amount of energy to the hadronic constituents in a given process. Specifically, it is seen from (10.24) that, energies being equal, strong processes require a *shorter* time interval to occur than electromagnetic ones. When the energy of the process increases, the time interval (10.24) taken by strong interactions falls off according to a hyperbolic law with respect to the time required for the electromagnetic interactions at the same energy. Indeed, in exchanging energy at "high" values, hadronic constituents look as bound particles to the e.m. probes, which see a "bond" with intensity greater than that produced by the e.m. interaction. Therefore, at increasing energy exchange, quarks appear more and more bound — that is, "confined". Thus quark confinement inside hadrons finds a natural (qualitative, at least) interpretation in the framework of DSR.

Then, due to the strong time deformation law (10.24), a hadronic system built up by strong bonds requires, in order to exchange energy among its constituents, a time interval which, with increasing energy, is still smaller than the time needed to supply energy to the system via e.m. interaction. Thus, the higher the supplied energy, the faster the bond responds to the solicitation.

The above considerations allow one to put in a different perspective the problem of isolating quarks. Such a problem is usually regarded as a "threshold energy" one, namely to find the energy value at which the system (a hadron, in this case) becomes unstable, so that its constituents might be isolated as particles "free" from the hadron system.

Apart from the question whether such an energy could be supplied to the system by means of a probe via its e.m. or strong interaction, the true problem is that one is just trying to let the hadron constituents to move in a space whose (Minkowskian) metric is not their own (strong) metric.

There is not yet a definite answer to such a question. But it is likely that, in the DSR framework, the proper way to put correctly the problem of the isolation of quarks is as follows: Getting an object to move in a space-time whose metric is not its own one (*i.e.* intrinsic to the interaction where it is acquainted to live in), and studying its motion. A possible analogy is provided by the motion of a real photon in presence of a gravitational field (*i.e.* in a space endowed with a gravitational metric) or by the motion of a virtual photon in a hadronic field.

CHAPTER 11

METRICS OF INTERACTIONS

The metric representation derived in the previous Chapters are, obviously, not ultimate. They constitute a mere phenomenological attempt at applying the formalism of the DSR (developed in Parts I and II), by showing explicitly how to pass from a deformed Minkowski metric to a metric representation of a given interaction.

The main aim of the present Chapter is just to provide a review of the phenomenological metrics derived for all four fundamental interactions, by stressing their general structure and their features as a premise for further developments.

11.1. Review of Phenomenological Metrics

11.1.1. *Energy-Dependent Phenomenological Metrics for the Four Interactions*

In Chaps. 7–10 it was shown that a local breakdown of Lorentz invariance may be envisaged for all four fundamental interactions (electromagnetic, weak, strong and gravitational) whereby one gets evidence for a departure of the space-time metric from the Minkowskian one (at least in the energy range examined). Let us review the basic features of the energy-dependent phenomenological metrics derived above:

(1) Both the electromagnetic and the weak metric show the same functional behavior, namely

$$\eta(E) = \text{diag}(1, -b^2(E), -b^2(E), -b^2(E)); \tag{11.1}$$

$$b^2(E) = \begin{cases} (E/E_0)^{1/3}, & 0 < E \leq E_0 \\ 1, & E_0 < E \end{cases} \tag{11.2}$$

$$= 1 + \theta(E_0 - E) \left[\left(\frac{E}{E_0} \right)^{1/3} - 1 \right], \quad E > 0 \tag{11.3}$$

(where $\theta(x)$ is the Heaviside theta function) with the only difference between them being the threshold energy E_0, *i.e.* the energy value at which the metric parameters are constant (namely the metric becomes Minkowskian: $\eta_{\mu\nu}(E \geq E_0) \equiv g_{\mu\nu} = \mathrm{diag}(1, -1, -1, -1)$); the fits to the experimental data yield

$$E_{0,\mathrm{e.m}} = (4.5 \pm 0.2)\,\mu\mathrm{eV}; \quad E_{0,\mathrm{weak}} = (80.4 \pm 0.2)\,\mathrm{GeV}. \tag{11.4}$$

Notice that for either interaction the metric is isochronous, spatially isotropic and subminkowskian, *i.e.* it approaches the Minkowskian limit from below (for $E < E_0$). Both metrics are therefore Minkowskian for $E > E_{0w} \simeq 80\,\mathrm{GeV}$, and then our formalism is fully consistent with electroweak unification, which occurs at an energy scale $\sim 100\,\mathrm{GeV}$.

(2) The strong metric reads

$$\eta_{\mathrm{strong}}(E) = \mathrm{diag}(b_{0,\mathrm{strong}}^2(E), -b_{1,\mathrm{strong}}^2(E),$$
$$- b_{2,\mathrm{strong}}^2(E), -b_{3,\mathrm{strong}}^2(E)); \tag{11.5}$$

$$b_{1,\mathrm{strong}}^2(E) = \left(\frac{\sqrt{2}}{5}\right)^2 ; \quad b_{2,\mathrm{strong}}^2(E) = \left(\frac{2}{5}\right)^2, \quad \forall E > 0 \tag{11.6}$$

$$b_{0,\mathrm{strong}}^2(E) = b_{3,\mathrm{strong}}^2(E) = \begin{cases} 1, & 0 < E \leq E_{0,\mathrm{strong}} \\ (E/E_{0,\mathrm{strong}})^2, & E_{0,\mathrm{strong}} < E \end{cases}$$
$$= 1 + \theta(E - E_{0,\mathrm{strong}}) \left[\left(\frac{E}{E_{0,\mathrm{strong}}}\right)^2 - 1 \right], \quad E > 0 \tag{11.7}$$

with

$$E_{0,\mathrm{strong}} = (367.5 \pm 0.4)\,\mathrm{GeV}. \tag{11.8}$$

In this case, contrarily to the electromagnetic and the weak ones, a deformation of the time coordinate occurs; moreover, the three-space is anisotropic, with two spatial parameters constant (but different in value) and the third one variable with energy like the time one.

(3) The explicit form of the gravitational energy-dependent metric is:

$$\eta_{\mathrm{grav}}(E) = \mathrm{diag}(b_{0,\mathrm{grav}}^2(E), -b_{1,\mathrm{grav}}^2(E), -b_{2,\mathrm{grav}}^2(E), -b_{3,\mathrm{grav}}^2(E)); \tag{11.9}$$

$$b^2_{0,\text{grav}}(E) = b^2_{3,\text{grav}}(E) = \begin{cases} 1, & 0 < E \leq E_{0,\text{grav}} \\ \dfrac{1}{4}(1 + E/E_{0,\text{grav}})^2, & E_{0,\text{grav}} < E \end{cases}$$

$$= 1 + \theta(E - E_{0,\text{grav}}) \left[\frac{1}{4} \left(1 + \frac{E}{E_{0,\text{grav}}} \right)^2 - 1 \right], \quad E > 0$$

$$(11.10)$$

(the coefficients $b^2_{1,\text{grav}}(E)$ and $b^2_{2,\text{grav}}(E)$ are presently undetermined at phenomenological level), with

$$E_{0,\text{grav}} = (20.2 \pm 0.1)\,\mu\text{eV}. \qquad (11.11)$$

It is worth stressing the analogy between the strong and the gravitational metrics. In both cases, a deformation of the time coordinate occurs. Moreover, one of the spatial parameters (we conventionally assumed as the third one) varies with energy as the time one in an overminkowskian way, namely, they approach the Minkowskian limit from above ($E_0 < E$). The other two spatial parameters are constant, but different in value for the strong case (*i.e.* the three-space is anisotropic for strong interactions), whereas nothing can be said for the gravitational case on an experimental basis.

The general pattern of the four phenomenological metrics is shown in Fig. 11.1.

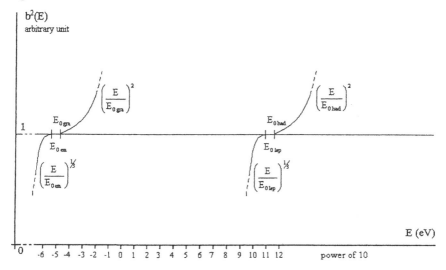

Fig. 11.1. General pattern of the phenomenological metrics for the four fundamental interactions.

11.1.2. *Threshold Energies and Recursive Metrics*

A comparison among the threshold energies for the electromagnetic, weak, strong and gravitational interactions, given by Eqs. (11.4), (11.9) and (11.12), yields

$$E_{0el} < E_{0grav} < E_{0w} < E_{0s} \qquad (11.12)$$

i.e. an increasing arrangement of E_0 from the electromagnetic to the strong interaction. Moreover

$$\frac{E_{0grav}}{E_{0el}} = 4.49 \pm 0.02; \qquad \frac{E_{0s}}{E_{0w}} = 4.57 \pm 0.01, \qquad (11.13)$$

namely

$$\frac{E_{0grav}}{E_{0el}} \simeq \frac{E_{0s}}{E_{0w}} \qquad (11.14)$$

an intriguing result indeed.

A further remark concerns the possible pattern of interactions ensuing from DSR. According to the results summarized above, we have two pairs of interactions: (i) electromagnetic and gravitational; (ii) weak and strong, ordered by the increasing arrangement of the threshold energies. Moreover, in each pair the former interaction is subminkowskian, and the latter is overminkowskian. The first question is: Does this pattern end with the second pair, or not? If a third pair exists, we can assume that the threshold energies of the new pair, E_{05} and E_{06}, are related to the threshold energies of the previous subminkowskian and overminkowskian metrics according to

$$\frac{E_{0,n+2}}{E_{0,n}} = \frac{E_{0,n+4}}{E_{0,n+2}}, \quad n = 1, 2 \qquad (11.15)$$

(with $E_{0el} = E_{01}$; $E_{0grav} = E_{02}$; $E_{0w} = E_{03}$; $E_{0s} = E_{04}$). In such hypothesis, with the values (11.4), (11.9) and (11.12) of the threshold energies for the known interactions, one gets

$$E_{05} \simeq 1.3 \times 10^{18} \text{ GeV};$$
$$E_{06} \simeq 6.7 \times 10^{18} \text{ GeV}. \qquad (11.16)$$

Let us also stress that, in the subminkowskian case, the deformed metrics do vanish, in general for an energy value $W_0 \neq 0$, *i.e.* by definition

$$b_\mu^2(W_0) = 0. \qquad (11.17)$$

The physical meaning of the metric vanishing is that, at the energy W_0, the interaction is unable to measure the space-time separation. In particular, such a value of the energy corresponds to a pointlike object for the interaction considered. Examples of this fact are provided by the inertial

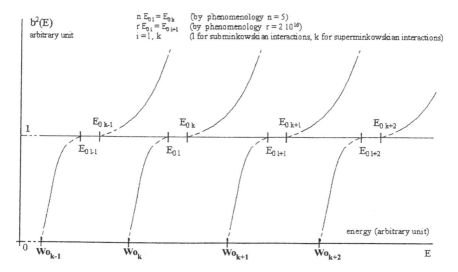

Fig. 11.2. Qualitative behavior of recursive and asymptotic metrics.

mass of the photon for an electromagnetic metric, or the neutrino masses for a leptonic metric. Indeed, for the e.m. interaction, $W_{0,\text{e.m.}}$ is the energy value E_{min} considered at the end of Sec. 7.1 (that was shown to be essentially the upper limit for the photon inertial mass).

A possible representation of the recursive pattern of the phenomenological metrics is illustrated in Fig. 11.2.

11.2. Asymptotic Metrics

Of course, the paucity of the phenomena, and the related experimental data, on which the phenomenological analysis is based, makes just preliminary and provisional the resulting metrics obtained for the electromagnetic, weak, gravitational and strong interactions.

In this connection, it is worth listing the possible functional forms of deformed metrics which might be able to describe physical interactions. We shall divide them in two main classes, according to the functional form of the parameters $b_\mu^2(E)$:

(I) First class

$$\begin{cases} b_0^2(E) = \text{const} \\ b_k^2(E) = \left(1 - \dfrac{W_{0k}}{E}\right)^{n_k} \end{cases} \tag{11.18}$$

where n_k is a certain real number.

These represent subminkowskian metrics, which are asymptotically minkowskian with increasing energy. Metrics of the first class are suitable for representing the electromagnetic and leptonic interactions.

(II) Second class

$$\begin{cases} b_0^2(E) = \left(1 + \dfrac{E}{W_{00}}\right)^{n_0} \\ b_k^2(E) = \left(1 + \dfrac{E}{W_{0k}}\right)^{n_k} \end{cases} . \tag{11.19}$$

Metrics of type II are obviously overminkowskian, divergent with increasing energy, asymptotically minkowskian with decreasing energy.

Metrics of the second class are suitable for the representation of the hadronic and the gravitational interaction.

From these results and those of Subsec. 11.1.2 on threshold energies, it is possible to illustrate on a qualitative basis the behavior of the asymptotic metrics, as done in Fig. 11.2.

Notice that both first-and second-class metrics become isotropic if the constants W_{0k}, n_k reduce to only two, W_0 and n. It is worth to stress some of the (remarkable) implications of the above metrics.

For both classes, the constants W_{0k} have the natural meaning of scale energy of the interaction described by the corresponding metrics. Moreover, in metrics of the first kind they represent the energy values at which the spatial part of the metric vanishes, in agreement with the definition given in Eq. (11.18).

Indeed, let us recall that, applying an isotropic metric of type I to the data discussed in Sec. 7.1 on e.m. wave propagation, one gets for $W_{0k,\text{e.m.}}$ values compatible with the present upper bounds on the photon inertial mass (since it is essentially the energy value $E_{\min}$ there defined).

As a last consequence, consider the general form of the time deformation law in a hadronic field, that reads:

$$\frac{\Delta t_{\text{had}}}{\Delta t_{\text{e.m.}}} = \frac{1}{\sqrt{\eta_{00}}} = \left(1 + \frac{E}{W_{00}}\right)^{-n_0/2} . \tag{11.20}$$

For $n_0 \le 2$ (on account of the results of the previous section) the above expression shows a behaviour asymptotically decreasing toward zero as the energy E increases,whereas, for $W_{00} \ge E > 0$, it goes to 1 as the energy decreases. Of course, such a behaviour is easily interpreted in terms of confinement and asymptotic freedom of hadronic constituents, according to the discussion at the end of Sec. 3.3.

Such a pattern may repeat itself again, or not (see Fig. 14), and it is of course a matter of experiment to check the real existence of these new pairs of interactions. What we exclude is that it repeats *ad infinitum*. In this connection, we recall that it was shown that the maximum possible force in Nature is provided by the *Kostro constant K*, given by[71]

$$K = \frac{c^4}{G} = 7.556 \times 10^{51} \, \text{GeV/cm} \qquad (11.21)$$

where G is the gravitational constant ($G = 1.072 \times 10^{-10} \, \text{cm}^5/[\text{GeV sec}^4]$). The corresponding maximum energy, *i.e.* the energy of the whole Universe, is therefore (assuming $R_0 \sim 10^{10}$ light-years)

$$E_{\max} = K R_0 \sim 10^{79} \, \text{GeV}. \qquad (11.22)$$

Either in the case of new interactions (besides the known ones) or not, we deem that the interaction pattern in the DSR scheme is bounded from above by the value $E_{\max}$ (11.22) related to the Kostro limit. This holds, in particular, for the asymptotic behaviour of the overminkowskian metrics.

PART IV

BREAKDOWN OF LOCAL LORENTZ INVARIANCE

CHAPTER 12

EXPERIMENTAL TESTS OF LOCAL
LORENTZ INVARIANCE

Let us now comment on the results obtained by the analysis of the experimental data for all the interactions considered. It is worth stressing that, in all four cases, one obtains evidence for a departure of the space-time metric from the minkowskian one. However, although a quite cautious attitude must be exercised, it would be probably wrong to understate the relevance of such findings. Let us therefore attempt to interpret them. The most straightforward meaning is that (neglecting the gravitational case, where of course a departure from the Minkowskian metric is expected and natural) the three phenomena considered (*i.e.* the superluminal propagation of evanescent e.m. waves in waveguide, the K_S^0 decay and the BE correlation) do show a breakdown of the usual Lorentz symmetry for the electromagnetic, weak and strong interaction, respectively. But — what's the basic point — one may wonder whether such a breakdown should be considered as an *actual* or an *effective* one. Namely, two interpretations can be given to the need for a deformed Minkowski metric in order to describe the above phenomena. The first one is to state that such results correspond to an actual local deformation of the space-time geometry, induced by the interaction considered. This would constitute an evidence in favor of a *real* analogy of the e.m., weak and strong interactions with the gravitational one, and therefore of the *real* validity of the solidarity principle for all the four interactions. The second possible interpretation is inspired by the analysis of the e.m. case. Indeed, in the e.m. wave propagation in waveguides, it is perhaps more sound to understand the non-minkowskian metric obtained in Sec. 3.1 as describing, in an effective way, the nonlocal e.m. effects which occur inside the waveguide and give rise to the superluminal propagation. In the same spirit, one can regard the arising of non-minkowskian metrics in the description of K_S^0 decay and BE correlation as due to the presence of nonlocal forces responsible for such phenomena. Otherwise speaking, the non-minkowskian metrics involved in such cases would play an *effective* role, in the sense that they would actually be a signature of the presence and effectiveness of nonpotential effects in the phenomena considered.

Any definite conclusion on the right significance to be attached to the possible evidence presented in Part III for the apparent breakdown of the Lorentz symmetry requires of course new experimental tests. It is however worth remarking that both interpretations of such findings have intriguing and far-reaching implications. Indeed, the open problems are: (i) Is there a *real* breakdown of the usual Lorentz symmetry in weak and strong interactions? (ii) Do such interactions require a description in terms of nonpotential forces? The answers to the above questions may obviously be obtained only by further investigations, from both the theoretical and the experimental side.

It is indeed an old-debated problem whether local Lorentz invariance (LLI) preserves its validity at any length or energy scale (far enough from the Planck scale, when quantum fluctuations are expected to come into play). Doubts on the reliability of a Lorentz-invariant description of physical phenomena at subnuclear distances were put forward, at the middle of sixties of the XX century, even in standard (and renowned) textbooks.[a]

From the experimental side, the main tests of LLI can be roughly divided in three groups[46]:

(a) Michelson–Morley (MM)-type experiments, aimed at testing isotropy of the round-trip speed of light;
(b) tests of the isotropy of the one-way speed of light (based on atomic spectroscopy and atomic timekeeping);
(c) Hughes–Drever-type (HD) experiments, testing the isotropy of nuclear energy levels.

All such experiments set upper limits on the degree of violation of LLI.

---

[a]Let us quote *e.g.* the words by Bjorken and Drell, who, in the introduction to their celebrated 1965 book "*Relativistic Quantum Fields*" (McGraw-Hill), explicitly write: "*There is nothing but positive evidence that special relativity is correct in the high-energy domain* (*i.e.* at "small" distances or in the "sub-microscopic" domain, authors' note), *and furthermore, there is, if anything, positive evidence that the notion of microscopic causality is a correct hypothesis. Since there exists no alternative theory which is any more convincing, we shall hereafter restrict ourselves to the formalism of local, causal fields. It is indoubtedly true that a modified theory must have local field theory as an appropriate large-distance approximation or correspondence. However, we again emphasize that the formalism we develop* (*i.e.* Relativistic Quantum Field Theory, authors' note) *may well describe only the large-distance limit* (*that is, distances* $>10^{-13}$ *cm*) *of a physical world of considerably different submicroscopic properties.*"

From the theoretical side, a lot of generalizations of Special Relativity and/or LLI breaking mechanisms exists in literature. Very interesting approaches to LLI breakdown within the framework of the Standard Model (SM) of fundamental interactions have been considered by Coleman and Glashow,[12] with the proposal of various tests of Special Relativity in cosmic-ray and neutrino physics, and by Jackiw,[72] who puts very stringent limits on such effects. Moreover, Kostelecky[73] built up a generalization of SM (*Standard-Model Extension*), allowing for violations of Lorentz and CPT symmetry that cannot occur in the usual SM, and providing a quantitative description of them in terms of a set of coefficients whose values are to be determined or constrained by experiment.

CHAPTER 13

A NEW ELECTROMAGNETIC TEST OF LLI

In the last decade of the XX century, Bartocci (together with Mamone-Capria)[74] proposed a new electromagnetic experiment aimed at testing LLI and able of providing direct evidence for its breakdown. The results obtained in a first, preliminary experimental run carried out in June 1998 — essentially aimed at providing new upper limits on the LLI breakdown parameter by an entirely new class of electromagnetic experiments — admit as the most natural interpretation the fact that local Lorentz invariance is in fact broken.[75,76]

The experiment was just repeated in 1999 with an improved apparatus. The analysis of this second run confirmed the positive evidence of the previous one.[77,78]

13.1. Theoretical Foundations

Let us consider a steady current I circulating in a closed loop γ, and a charge q subjected to the uniform magnetic field produced by γ.

If both q and γ are at rest in the same reference frame K in which electromagnetism can be described by standard Maxwell theory, then no force $\mathbf{F}$ acts on q due to γ. In fact, the expression of this force is:

$$\mathbf{F} = q \left(-\nabla \Phi - \frac{\partial \mathbf{A}}{\partial t} + \mathbf{V} \times \mathbf{B} \right), \tag{13.1}$$

(where Φ and $\mathbf{A}$ are respectively the electric and magnetic potential associated to γ, $\mathbf{B} = \operatorname{curl} \mathbf{A}$ the corresponding magnetic field, and $\mathbf{V}$ the velocity of q in K) and, in the aforesaid conditions, $\Phi = 0$, $\frac{\partial \mathbf{A}}{\partial t} = 0$, $\mathbf{V} = 0$. Notice that, as a matter of fact, the equation $\Phi = 0$ is just an additional hypothesis in standard electrodynamics, the so-called *Clausius postulate*, which should hold of course just for ideal one-dimensional circuits. For real ones, the validity of this equation has been questioned both on an experimental basis[79] and on a theoretical one.[80,81]

Let us now introduce a reference frame K' moving with some uniform velocity $\mathbf{v}$ with respect to K. Then, by means of the Principle of Relativity, one can assert that there is no force acting on q, if both q and γ are at rest in this new frame K'.

The point is: Is there some (experimental and/or theoretical) possibility that some force does act on q in this case?

Such a problem was faced by Bartocci[74] in the search for a discriminating test between "classical" and "relativistic" electrodynamics. Let us briefly summarize their results.

In order to compute the required force without principle of relativity, we can use again Eq. (13.1), where of course $\mathbf{V}$ becomes equal to $\mathbf{v}$, and we suppose that $\Phi = 0$, an hypothesis which makes all the difference with the relativistic interpretation of Maxwell electrodynamics, precisely that difference we want just to test.

Let us call x, y, z, t the space-time coordinates in K, and suppose that:

(a) the velocity $\mathbf{v}$ is equal to the constant velocity $(v, 0, 0)$;
(b) the "moving" circuit is parametrized by the equations $x = r\cos\theta + vt$, $y = r\sin\theta$, $z = 0$;
(c) the charge q is placed in the point $P = (vt, y, 0)$.

Then, cumbersome but straightforward calculations yield the expression (approximated up to terms of second order in $\beta = v/c$):

$$
\mathbf{F} = -\frac{\mu_0 q I v}{4\pi r}\left[1 + \left(\frac{y}{r}\right)^2\right]^{-3/2} \left\{ \int_0^{2\pi} \frac{\sin^2\theta - \dfrac{y}{r}\sin\theta}{\left[1 - \dfrac{\dfrac{2y}{r}}{1 + \left(\dfrac{y}{r}\right)^2}\sin\theta\right]^{3/2}} d\theta \right\} \mathbf{e}_y,
$$

(13.2)

with $\mathbf{e}_y$ being the unit vector in the y-direction.

Instead of a single charge, let us consider a tiny conductor R of length $L < r$, lying on the (x, y)-plane, with an end in the center O of the loop. Then, in order to find the voltage $V(s)$ across the conductor, one has simply to integrate the field corresponding to Eq. (13.2) with respect to y, thus getting

$$
V(s) = \frac{\mu_0 I v}{4\pi} \int_0^s [1 + u^2]^{-3/2} \left\{ \int_0^{2\pi} \frac{u \sin\theta - \sin^2\theta}{\left(1 - \dfrac{2u}{1 + u^2} \sin\theta\right)^{3/2}} \, d\theta \right\} du
$$
$$
\approx 10^{-7} I v i(s), \tag{13.3}
$$

where we put $s = L/r < 1$, and $i(s)$ is the double integral

$$
i(s) = \int_0^s [1 + u^2]^{-3/2} \left\{ \int_0^{2\pi} \frac{u \sin\theta - \sin^2\theta}{\left(1 - \dfrac{2u}{1 + u^2} \sin\theta\right)^{3/2}} \, d\theta \right\} du, \tag{13.4}
$$

which can be numerically evaluated for different values of s.

Of course, the existence of the above voltage is strictly related to the relative motion of the conductor with respect to the reference frame K, or, better, with respect to the magnetic field **B**. Therefore, a measurement of a $V \neq 0$ for a system of a coil and a conductor, seemingly at rest in the laboratory frame, could constitute a direct evidence of the breakdown of local Lorentz invariance. Such a breakdown could be parametrized in terms of a speed v trough Eq. (13.3), and one of its possible interpretations (as we shall discuss later on in more details) is assuming that actually the magnetic field produced by the coil is not at rest with it, but — due to a kind of "kinematical decoupling" — is at rest with respect to some reference frame (just moving with respect to Earth with speed v) which drags it.

It is worth stressing that such an effect does not depend on the magnitude of the speed v as compared to the light speed c, and one may be able to observe it even for very small v by suitably increasing I. One would have therefore, in this case, a breakdown of LLI at low speeds, contrarily to what is generally believed.

Moreover, it is expected, on the basis of the above discussion, that this effect critically depends on the orientation of the plane where the system lies. This is reflected in the structure of the experimental apparatus devised to looking for it, as we shall see in the next section.

13.2. Experimental Setup and Measurement Procedure

The new test first proposed by Bartocci is based on the possibility of detecting a non-zero Lorentz force between the magnetic field **B** generated by a stationary current I circulating in a closed loop γ, and a charge q, in the hypothesis that both q and γ are at rest in the same inertial reference frame. Such a force is zero, according to the standard (relativistic) electrodynamics.

The experimental setup was devised in order to put new upper limits on the breakdown of LLI, by means of such an entirely new class of electromagnetic experiments, and also to test possible anisotropic effects in such limits.

The experimental device used in both experiments is schematically depicted in Fig. 13.1, whereas the electric schemes of the apparatus and the measurement device are sketched in Figs. 13.2 and 13.3.[a] It consisted of a Helmholtz coil γ and a Cu conductor R placed inside it on a plane orthogonal to the γ axis. The conductor R was connected in series to a

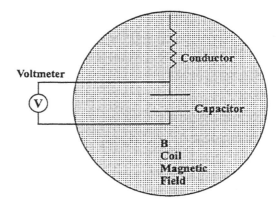

Fig. 13.1.   Schematic view of the experimental setup.

---

[a]We have to warn the reader against possible misunderstandings of Fig. 13.1. In the picture the voltmeter cables appear to be orthogonal to the coil axis, and therefore to the magnetic field **B**. Actually, this is due to the fact that the picture is two-dimensional, and not three-dimensional. In fact, in the experimental device the voltmeter cables were placed parallel to the coil axis (and therefore to **B**), just in order to avoid spurious induction currents (see Figs. 13.2 and 13.3).

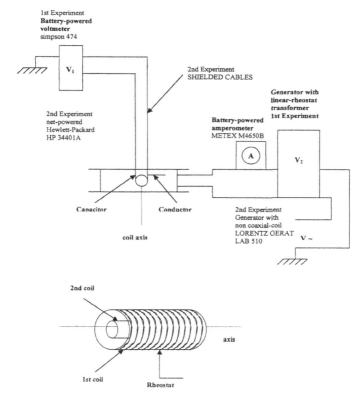

Fig. 13.2.   Schematic front view of the Helmholtz coil and the capacitor-resistor probe.

capacitor C,[b] and a voltmeter was connected in parallel to the capacitor, so to measure the voltage due to a possible gradient of charge across R. The conductor could change its orientation in the coil plane. Moreover, the whole system of the RC circuit and the coil could turn so letting its plane coincide with one of the coordinate planes.

The center of the geometrical coordinate system coincided with the center of the coil. The coordinate system was chosen as follows: the (x, y) plane tangent to the Earth surface, with the y-axis directed as the (local) Earth magnetic field $\mathbf{B}_T$; the z-axis directed as the outgoing normal to the Earth surface, and the x-axis directed so that the coordinate system is

---

[b]The value of the capacitance C of C was $1pF$ in both experiments. The capacitance of the conductor R was calculated in both cases and found to be about two orders of magnitude lower with respect to C, and therefore fully negligible because the capacitance of R is in parallel to C.

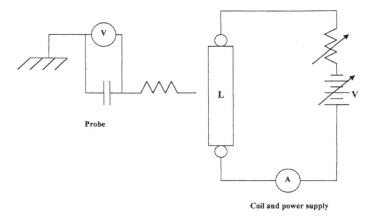

Fig. 13.3. Sketch of the electric circuits of the capacitor-resistor probe and the Helmholtz coil with power supply.

left-handed. The conductor orientation in the plain coil was parametrized in terms of an angle α (ranging from 0 to 2π). The rotation of α was chosen clockwise in the plain coil with respect to an observer oriented along the coordinate axis orthogonal to the coil plane. The first orientation of R corresponding to the angle $\alpha = 0$ was along the negative direction of the z-axis in the case of the two vertical canonical planes (see Fig. 13.4 for the (x, z)-plane), and along the negative direction of the y-axis in the case of the horizontal plane. A steady-state current I circulating in the coil produced a constant magnetic field $\mathbf{B}$ in which the RC circuit is embedded. The circuit and the coil were mutually at rest in the laboratory frame.

13.3. Experimental Results

13.3.1. *First Experiment*

In the first experiment, the measurements performed with the system lying on the planes (x, y) and (y, z) gave values of V compatible with the instrument zero. Indeed, in such cases the statistical tests of correlation showed that each of the points outside the zero-voltage band is uncorrelated with the preceding and the subsequent point either, and the whole set of points was shown to be uncorrelated ($R^2 < 30\%$). Let us recall that each point is the average of five measurements, taken at the same angle. As to the measurements in the plane (x, z), it was shown instead that the four points outside the zero band are statistically correlated ($R^2 > 80\%$), and so they represent a valid candidate for a nonzero signal.

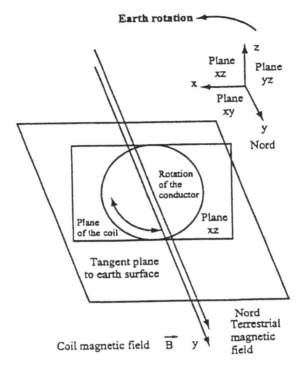

Fig. 13.4.   Schematic view of the orientation of the apparatus and the related magnetic fields in the (x, z) plane.

A polynomial interpolating curve for these points is shown in Fig. 13.5. Such an interpolating procedure was essentially aimed at finding the angle α_{max} corresponding to the maximum value of V, $V^{xz}_{max} = (36.0 \pm 1.0)\,\mu V$. The value found was $\alpha_{max} = 3.757\,\text{rad}$. The knowledge of α_{max} is needed in order to determine the value of the anisotropic LLI violation parameter (see Subsec. 13.4).

For comparison, the results of the first experiment are given in Table 13.1.

13.3.2. *Second Experiment*

In analogy with the 1998 experiment, the adopted criterion in order to accept a voltage peak in a given plane was the presence of at least three points statistically correlated with maximum value of voltage at least 2σ from the other (uncorrelated) points. The corresponding voltage value will

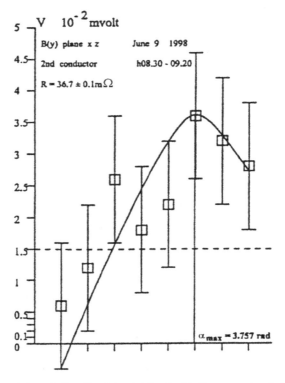

Fig. 13.5.   Curve interpolating the data obtained with the apparatus in the (x, z) plane, showing the angle of maximum signal $\alpha_{\max} = 3.757\,\text{rad}$.

Table 13.1   Results of the 1998 experiment for the measured V. Symbol: V_{am} = average of a.m. measurements

	$B = 3.65\,\text{mT}$
plane	V (in μV)
xy	$V_{\text{am}} = 17.4 \pm 0.3$
yz	$V_{\text{am}} = 15.9 \pm 0.2$
xz	36.0 ± 1.0
	$V_0 = (15.0 \pm 10.0)\mu\text{V}$

be referred to as a *peak value*. In absence of peaks, one calculates the average of all the measured values outside the instrumental zero (*level value*).

In the second experiment, a signal candidate (in the form of a peak signal) was analogously found in the plane (x, z). For

$B = B_1 = (5.14 \pm 0.01)$mT, the average peak value was in excellent agreement with the result of the former experiment: $V_{max}^{'xz} = 35.4 \pm 0.1\,\mu$V. The signal was again highly anisotropic, and its behaviour with α is the same as depicted in Fig. 13.5. However, now possible signal candidates were found also in the planes (x, y) and (y, z) (this is also a consequence of the higher sensitivity of the multimeter, improved by two orders of magnitude). In those planes, there was no dependence on α, and therefore no spatial anisotropy. On the contrary, a time anisotropy was found in the (x, y) plane, since the measurements taken a.m. gave values within the instrument zero. The average level values found for $B = B_1$ were $V^{'xy} = (30.7 \pm 0.1)\mu$V and $V^{'yz} = (26.6 \pm 0.1)\mu$V. The measurements taken with the halved value of the coil magnetic field, $B = B_2 = (2.58 \pm 0.01)$mT, gave similar results, with voltage values $V_{max}^{'xz} = (41.8 \pm 0.1)\mu$V, $V^{'xy} = (34.4 \pm 0.1)\mu$V and $V^{'yz} = (30.6 \pm 0.1)\mu$V. Not only these values were not halved with respect to those obtained for $B = B_1$ (as expected in the case of a linear relation between V and I, like that derived via the Lorentz force), but surprisingly enough they were slightly higher! Moreover, a check was made by reversing the coil current. No change in the sign of V occurred. This allows one to conclude that the observed effect is independent of the direction of the current (and depends apparently in the "wrong way" on its magnitude).

The results obtained in the second experiment are summarized in Table 13.2.

In the case of the (x, y)-plane, we gave no value of $\langle V \rangle$ because the value of V_{am} was within the instrumental zero.

Table 13.2   Results of the 1999 experiment for the measured V. Symbols: $V_{am(pm)}$ = average of a.m. (p.m.) measurements; $\langle V \rangle = (V_{am} + V_{pm})/2$

plane	V_{am} (μV)	V_{pm} (μV)	$\langle V \rangle$ (μV)	isotropy	
		$B = B_1 = 5.14$ mT		space	time
xy	17.7 ± 1.6	30.7 ± 1.5		yes	no
yz	24.1 ± 2.4	29.1 ± 2.1	26.5 ± 1.8	yes	yes
xz	35.3 ± 0.5	37.0 ± 0.5	36.2 ± 0.7	no	yes
		$B = B_2 = 2.58$ mT			
xy	16.8 ± 1.6	34.4 ± 1.5		yes	no
yz	27.0 ± 1.7	34.1 ± 1.4	30.6 ± 1.7	yes	yes
xz	43.8 ± 0.3	39.7 ± 0.3	41.8 ± 0.7	no	yes
V_0	16.5 ± 1.5			yes	

13.3.3. An Interpretation by DSR

An attempt at interpreting the above experimental results can be given in terms of the phenomenological metrics derived in the framework of DSR.

To this aim, let us first discuss in more detail the implications of the spatial anisotropy of the observed effect.

Define the quantities

$$D \equiv V_{am}^{xz} - V_{am}^{xy}; \qquad (13.5)$$

$$\Delta \equiv \langle V^{xz} \rangle - V_{am}^{xy}. \qquad (13.6)$$

The values of D, Δ are given in Table 13.3. Notice that such values increase with decreasing B. This fact will be discussed later on.

The averages $\langle D \rangle$, $\langle \Delta \rangle$ are therefore

$$\langle D \rangle = (21.1 \pm 0.9)\mu V; \quad \langle \Delta \rangle = (20.7 \pm 1.0)\mu V. \qquad (13.7)$$

Let us attempt an interpretation of these results (although preliminary). Both experiments show a (peak) signal in the (x, z)-plane. Such an effect could, at a first sight, be attributed to a gravitational breakdown of the local Lorentz invariance. However, the results obtained in the second experiment, which show a (level) signal in the plane (x, y), suggest that the observed effect is both gravitational and electromagnetic. If so, then it is possible to conclude that the difference between the measured voltages in the two planes (namely, the quantities $\langle D \rangle$, $\langle \Delta \rangle$) provides a measure of the amount of the gravitational contribution to the measured voltage versus the electromagnetic one.

This is indeed supported by the agreement of the values (12.3) with the threshold energy for the gravitational metric $E_{0,grav}$ (Eq. (11.12)).

On the other hand, let us evaluate the differences in the voltages measured in the three planes. According to the second experiment with $B = B_1$ (cf. Table 13.3), the transitions of the LLI breakdown value of the voltage.

(i) from the space anisotropy in the plane (x, z) ($\langle V^{xz} \rangle = (36.2 \pm 0.7)\,\mu V$) to the time anisotropy in the plane (x, y) ($V_{pm}^{xy} = (30.7 \pm 1.5)\,\mu V$), and

Table 13.3

B (mT)	D (μV)	Δ (μV)
$B_1 = 5.14$ (I exp.)	17.7	18.5
$B = 3.65$ (II exp.)	18.6	18.6
$B_1 = 2.58$ (I exp.)	27.0	25.0

(ii) from the time anisotropy in the plane (x, y) to the space-time isotropy in the plane (y, z) ($\langle V^{yz} \rangle = (26.5 \pm 1.8)\,\mu\text{V}$), do occur for steps $\langle \Delta V \rangle \simeq 4.85\,\mu\text{V}$:

$$\langle V^{xz} \rangle - V^{xy}_{\text{pm}} = 5.5\,\mu\text{V}; \quad V^{xy}_{\text{pm}} - \langle V^{yz} \rangle = 4.1\,\mu\text{V}. \tag{13.8}$$

This does just agree with the value (11.4) of the threshold energy for the electromagnetic metric $E_{0,\text{em}}$. In general, we expect that the transitions can occur for multiples (by relative integers) of $\Delta V \sim 5\,\mu\text{V}$ (a sort of *"quantization of metric variation"*).

The fact that both electromagnetic and gravitational metrics must be taken into account in order to interpret the results of this type of experiments (namely, either interaction is responsible of the observed LLI breakdown) is suggested also by the value of the instrumental zero V_0 in both experiments, $V_0 \propto (E_{0,\text{grav}} - E_{0,\text{e.m.}})$, the proportionality constant being obviously the electron charge e.[c]

In this context, it is worth noticing two intriguing features of V_0. On one side, the energy value corresponding to V_0 lies inside the energy interval $(E_{0\text{el}}, E_{0\text{grav}})$ where either metric is exactly Minkowskian; moreover, such a value coincides with the amplitude of that interval of Minkowskian behavior. This circumstance seems to provide also an explanation of the apparent independence of the instrumental zero from the measuring device.

We can therefore conclude that, even if the first experiment left open the possibility of a mere gravitational explanation of the observed effect, the results of the second one are strongly in favor also of an inescapable electromagnetic contribution to the breakdown of the local Lorentz invariance.

Let us attempt also a qualitative explanation of the strange fact (cf. Subsec. 13.3.2 and Table 13.2) that one finds voltage values slightly higher for the halved value B_2 of the magnetic field in the second experiment. It is indeed $V^{xy}_{\text{am}}(B_1) \simeq V^{xy}_{\text{am}}(B_2)$ but $V^{xy}_{\text{pm}}(B_2) > V^{xy}_{\text{pm}}(B_1)$; $\langle V^{yz}(B_2) \rangle >$

---

[c]The (relative) difference in sign between the gravitational and electromagnetic contributions to V_0 is essentially due to the opposite behavior of the two corresponding metrics (see Subsec. 11.1). In fact, the gravitational metric is over-Minkowskian, and reaches the limit of Minkowskian metric for decreasing values of E (with $E > E_{0,\text{grav}}$), whereas the electromagnetic metric is sub-Minkowskian and thus attains the Minkowskian limit for increasing values of the energy ($E < E_{0,\text{e.m.}}$). Therefore the two metrics become Minkowskian for energy variations of opposite sign.

$\langle V^{yz}(B_1)\rangle$; $\langle V^{xz}(B_2)\rangle > \langle V^{xz}(B_1)\rangle$. In particular, one has

$$d^{xy} \equiv V^{xy}_{\text{pm}}(B_2) - V^{xy}_{\text{pm}}(B_1) = 3.7\,\mu\text{V};$$
$$d^{yz} \equiv \langle V^{yz}(B_2)\rangle - \langle V^{yz}(B_1)\rangle = 4.1\,\mu\text{V}; \qquad (13.9)$$
$$d^{xz} \equiv \langle V^{xz}(B_2)\rangle - \langle V^{xz}(B_1)\rangle = 5.6\,\mu\text{V}$$

and therefore

$$\langle d\rangle = 4.5\,\mu\text{V} \qquad (13.10)$$

again in agreement with the value of $E_{0,\text{e.m.}}$. This effect can be therefore interpreted as an electromagnetic one. When the voltage values are within the instrumental zero V_0 (the case of V^{xy}_{am}), we are in a region where both metrics are Minkowskian and there is then no difference between the two values corresponding to B_1 and B_2. For $V > V_0$, decreasing the value of the magnetic field amounts to decrease the energy of the system. According to the subminkowskian behaviour of the electromagnetic metric, this implies a stronger departure from the Minkowski metric, i.e. an enhanced effect of breakdown of the Lorentz invariance which manifests itself in a higher value of the voltage (contrarily to what expected on classical electrodynamic arguments based e.g. on the Lorentz force). This point is further enforced by the values of the quantities D and Δ, Eqs. (13.5)–(13.6) (increasing with decreasing B: see Table 13.3).

13.4. LLI Violation Parameter

We want now to give an estimate of the amount of breakdown of LLI ensuing from our experiment, in order to compare it with the existing limits.[46]

Let us introduce a *phenomenological LLI breakdown speed* v, defined as the relative speed between the coil γ and the conductor R, needed to provide, by means of the Lorentz force $\mathbf{F} = q\mathbf{v} \times \mathbf{B}$, the maximum measured voltage V_{max} across R. A physical meaning can be given to such a speed, if one assumes (as a possible interpretation) that the observed effect is due to a kinematical decoupling of the magnetic field $\mathbf{B}$ from the coil that generates it. As a consequence, the coil and the conductor are at rest in the same frame (the laboratory frame), whereas the field $\mathbf{B}$ is at rest with respect to an absolute reference frame Σ_0. Possible candidates for Σ_0 are: (a) The frame where the $2.7^\circ K$ background thermal radiation is isotropic for all the velocities of light; (b) the Hubble frame, where an observer would see all galaxies receding away with the Hubble expansion velocity; (c) the frame tied to the moving arm of our Galaxy; (d) the frame of the stochastic background gravitational radiation.

We can get an estimate of such a speed v (the Earth speed with respect to the absolute reference frame Σ_0, in the kinematical decoupling interpretation) by means of Eqs. (13.3) and (13.4). For $s = L/r = 12.00/13.25 \simeq 0.9$, a numerical evaluation of the integral in Eq. (13.4) gives approximately the value $i(0.9) \approx 5.08$. With $V = V_{\max} = (3.6 \pm 1.0) \times 10^{-5}$ volt, $I = 5A$, Eq. (13.3) yields therefore

$$v = (5.906 \pm 0.001) \times 10^{-2}\, \mathrm{m/sec}\,. \tag{13.11}$$

It is now easy to see why it is impossible to detect such an effect by means of an experiment of the Michelson–Morley-type. As is well known, the displacement Δn of the interference fringe in a MM experiment is given by

$$\Delta n = (\ell_1 + \ell_2)\frac{1}{\lambda}\left(\frac{v_R}{c}\right)^2, \tag{13.12}$$

where ℓ_1, ℓ_2 are the length of the arms of the interferometer, λ the light wavelength, and $v_R \simeq 3 \times 10^4\, \mathrm{m/sec}$ the Earth revolution velocity. In the original MM experiment, it is $\ell_1 + \ell_2 = 22\,\mathrm{m}$, $\lambda = 5.5 \times 10^{-7}\,\mathrm{m}$, $\Delta n = 0.4$. In our case, we have to replace v_R by the effective LLI breakdown speed v, whose value, according to the experimental findings (and the proposed interpretation), is given by Eq. (13.11) ($v \simeq 0.06\,\mathrm{m/sec}$). Then, by using the same parameters of the original MM experiment, one gets

$$\Delta n \simeq 0.2 \times 10^{-11}, \tag{13.13}$$

a fringe displacement completely unobservable even by modern tools.

We recall that two different kinds of LLI violation parameters exist: the isotropic (essentially obtained by means of experiments based on the propagation of e.m. waves, e.g. of the Michelson–Morley type), and the anisotropic ones (obtained by experiments of the Hughes–Drever type,[46] which test the isotropy of the nuclear levels).

In the former case, the LLI violation parameter reads[46]

$$\delta = \left(\frac{u}{c}\right)^2 - 1 \tag{13.14}$$

$$u = c + v$$

where c is, as usual, the speed of light in vacuo, v is the LLI breakdown speed (e.g. the speed of the preferred frame) and u is the new speed of light (which is nothing but the maximal causal speed of the electromagnetic interaction, in DSR, or the "maximum attainable speed", in the words of Coleman and Glashow[12]). The smallest upper limit obtained in this case is[46]

$$\delta < 10^{-8}. \tag{13.15}$$

In the anisotropic case, there are different contributions δ^A to the anisotropy parameter from the different interactions. In the HD experiment, it is A = S, HF, ES, W, meaning strong, hyperfine, electrostatic and weak, respectively. These correspond to four parameters δ^S (due to the strong interaction), δ^{ES} (related to the nuclear electrostatic energy), δ^{HF} (coming from the hyperfine interaction between the nuclear spins and the applied external magnetic field) and δ^W (the weak interaction contribution). The upper limits on the anisotropic parameter range from

$$\delta < 10^{-18} \tag{13.16}$$

of the HD experiment to

$$\delta < 10^{-27} \tag{13.17}$$

of the Washington experiment.[46] Let us also report the upper limit derived by Coleman and Glashow,[12] that is

$$\delta < 10^{-23} \tag{13.18}$$

(which is the same obtained for δ^S, the anisotropic parameter due to the strong interaction, in the HD experiment).

Although the observed effect is strongly anisotropic, let us assimilate it to an isotropic one and consider first the latter case. We have to use Eq. (13.14), with the value (13.11) of the LLI breakdown speed. Then, one gets

$$\delta = \left(\frac{u}{c}\right)^2 - 1 \simeq 2\frac{v}{c} = 3.94 \times 10^{-11} \tag{13.19}$$

which is lower by two orders of magnitude than the upper limit (13.15) for the isotropic case.

Let us now consider the more realistic anisotropic case. To calculate the anisotropic parameter, we exploit the formula (derived by the so-called $TH\varepsilon\mu$ formalism[46])

$$\delta = A\left[1 - \left(\frac{u}{c}\right)^2\right](v_\perp^2 - 2v_\parallel^2) \tag{13.20}$$

where u is the maximal causal speed, $v_\perp$, $v_\parallel$ are the components of the LLI breakdown velocity $\mathbf{v}$ orthogonal and parallel to the applied magnetic field **B**, respectively, and A is a constant which depends on the actual anisotropic experiment considered. For the HD case, it is $A = \frac{2}{15}$. The use of this formula is suggested, among the others, by the analogous role played, in the coil experiment and in the HD one, by the magnetic field as fixing a preferred direction in space. For u, one can take the two maximal speeds obtained in the hadronic case from the analysis of the Bose–Einstein correlation, namely (see Eq. (10.22))

$$u_1 = \frac{5}{2}c; \quad u_2 = \frac{5}{\sqrt{2}}c. \tag{13.21}$$

Since the relative orientation between **B** and **v** during the experiments (in correspondence to a given angle α) is not known, it is only possible to find the extrema of the values taken by $\delta(u)$. Thus, the maximum value $\alpha_{\max}$ is to be considered in Eq. (13.20).

One has therefore

$$\delta_i^{1,2} \equiv \delta^{1,2}(u_i, \alpha) = \frac{2}{15}\left[1 - \left(\frac{u_i}{c}\right)^2\right] v^2 \times \begin{cases} (\cos^2 \alpha - 2\sin^2 \alpha), \\ (\sin^2 \alpha - 2\cos^2 \alpha) \end{cases}, \quad i = 1, 2 \tag{13.22}$$

and finally

$$\delta_1^{1,2} = \begin{cases} 1.9 \times 10^{-29} \\ 2.7 \times 10^{-20} \end{cases} \tag{13.23}$$

$$\delta_2^{1,2} = \begin{cases} 4.2 \times 10^{-29} \\ 5.9 \times 10^{-20} \end{cases}. \tag{13.24}$$

Then, the anisotropic parameter corresponding to the e.m. LLI breaking effect is in the range

$$1.9 \times 10^{-29} < \delta < 5.9 \times 10^{-20} \tag{13.25}$$

and therefore compatible with the upper limits (13.16), (13.17).

However — at the light of the discussion of Sec. 13.3 — the parameters (13.19), (13.25) have to be regarded actually as the result of two contributions to the LLI breaking effect, one of gravitational and the other of electromagnetic nature, whereas the existing limits refer to the e.m. interaction (light propagation experiments) and e.m., leptonic and hadronic interactions (nuclear energy level measurements).

The new experimental situation of the LLI parameters is summarized in Fig. 13.6.

We want to stress that, in general, in the usual analysis of the LLI violating parameters, one looks for a LLI breakdown speed v which is in a sense

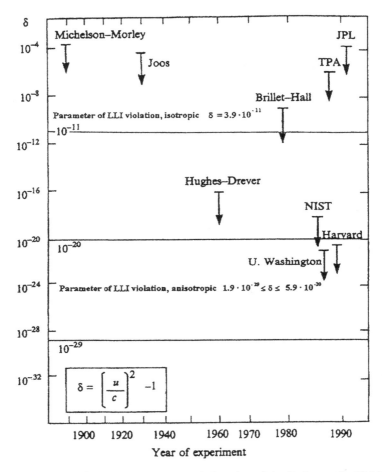

Fig. 13.6.   Showing the present experimental situation of the limits on the LLI break-down parameter δ (adapted from Will, Ref. [46], p. 322). The three horizontal straight lines are the limits obtained in the present experiments.

external to the interaction ruling the physical system under measurement. Indeed, typical candidates for v are the Earth revolution speed around the Sun ($v = 30$ Km/s) or the drift speed of the solar system in the Galaxy ($v = 300$ Km/s). In the present case, one is assuming that the LLI breaking speed is actually *internal* to the system. It parametrizes the amount of LLI breakdown inherent to (and characteristic of) the interaction ruling the process. This different point of view is typical of the DSR formalism, and is reflected in the use we made of Eq. (13.14). Indeed, in order to calculate δ

in the anisotropic case, one exploits the values of the maximal causal speeds derived from the analysis of the strong interaction based on BE correlation (the only interaction at present described by a spatially anisotropic metric) (cf. Chap. 10).

13.5. Appendix — Einstein and the Aether

When Einstein started his scientific life as a young physicist, the aether was conceived as a material medium allowing propagation of the electromagnetic signals and filling the interplanetary and interstellar spaces. This was for instance the idea of Maxwell, as formulated by him in the article he wrote in the *Encyclopedia Britannica*. According to Drude and Cohn, the aether might also be thought as the vacuum, endowed with suitable physical properties.

Einstein was so convinced about the reality of the aether, that he devised as early as 1895 an experiment aimed to detect the drift of the aether. He pursued such an idea when a student at Zurich ETH in 1897. However, he was not allowed to carry out such an experiment by his professor H. F. Weber. Later, he sent to his uncle C. Koch a manuscript (remained unpublished) entitled *On the Examination of the State of the Aether in a Magnetic field*. In the experiment therein proposed, he wanted to investigate the reciprocal effect between the magnetic field generated by a steady current and the aether, regarded as an elastic medium.

Another experimental proposal was exposed by Einstein in a letter to M. Grossmann from Winterthur in 1901. Subsequently, Einstein abandoned the idea of the aether in the old sense, and rather identified it, within General Relativity, as a substratum without mechanical and kinematical properties, but able to codetermine mechanical and electromagnetic events. In a modern view, we can state that such a new concept of aether is nothing but a static gravitational field (locally the Minkowski space-time).[82]

The possible existence of the aether is indeed nothing, in today language, than the possibility of LLI breakdown. We can therefore conclude that Einstein asked the right question, but looking for the wrong answer. His experiment based on the feedback between a steady current and the aether has strong analogies with Bartocci's one we discussed above. In particular, it belongs to the same class of LLI breakdown tests, namely those based on electromagnetic effects.

CHAPTER 14

THE GRAVITATIONAL MASS OF ELECTRON FROM GEOMETRY

In this Chapter, it will be shown that the DSR formalism yields an expression of the electron mass m_e in terms of the parameter δ of local Lorentz invariance (LLI) breakdown and of the threshold energy for the gravitational metric, $E_{0,\text{grav}}$ (*i.e.* as by now familiar, the energy value under which the metric becomes Minkowskian). This allows one to evaluate m_e from the (experimental) knowledge of such parameters.[83]

14.1. LLI Breaking Factor and Relativistic Energy in DSR

The breakdown of standard local Lorentz invariance (LLI) is expressed by the LLI breaking factor parameter δ discussed in Sec. 13.4. In particular, the isotropic parameter is expressed by Eq. (13.14). To the present aims, one can also define δ as follows:

$$\delta_{\text{int}} \equiv \frac{m_{\text{in,int}} - m_{\text{in,grav}}}{m_{\text{in,int}}} = 1 - \frac{m_{\text{in,grav}}}{m_{\text{in,int}}} \qquad (14.1)$$

where $m_{\text{in,int}}$ is the inertial mass of the particle considered with respect to the given interaction.[a] In other words, we assume that the *local* deformation of space-time corresponding to the interaction considered, and described by the metric (3.3), gives rise to a *local violation* of the Principle of Equivalence for interactions different from the gravitational one. Such a departure, just expressed by the parameter δ_{int}, does constitute also a measure of the amount of LLI breakdown. In the framework of DSR, δ_{int} embodies the geometrical contribution to the inertial mass, thus discriminating between two different metric structures of space-time.

---

[a]In the following, "int" denotes a physically detectable fundamental interaction, which can be operationally defined by means of a phenomenological energy-dependent metric of deformed minkowskian type.

Of course, if the interaction considered is the gravitational one, the Principle of Equivalence strictly holds, *i.e.*

$$m_{\text{in,grav}} = m_g \tag{14.2}$$

where m_g is the gravitational mass of the physical object considered, *i.e.* it is its "gravitational charge" (namely its coupling constant to the gravitational field).

Then, (14.1) can be rewritten as:

$$\delta_{\text{int}} \equiv \frac{m_{\text{in,int}} - m_g}{m_{\text{in,int}}} = 1 - \frac{m_g}{m_{\text{in,int}}} \tag{14.3}$$

and therefore, when the particle is subjected only to gravitational interaction, it is

$$\delta_{\text{grav}} = 0. \tag{14.4}$$

Let us recall the expression (5.10) of the deformed relativistic energy, for a particle subjected to a given interaction and moving along $\hat{x}^i$:

$$
\begin{aligned}
E_{\text{int}} &= m_{\text{in,int}} u_{i,\text{int}}^2(E) \tilde{\gamma}_{\text{int}}(E) \\
&= m_{\text{in,int}} c^2 \frac{b_{0,\text{int}}^2(E)}{b_{i,\text{int}}^2(E)} \left[1 - \left(\frac{v_i b_{i,\text{int}}(E)}{c b_{0,\text{int}}(E)} \right)^2 \right]^{-1/2}
\end{aligned}
\tag{14.5}
$$

(with $\mathbf{u}_{\text{int}}(E)$ the maximal causal velocity for the interaction considered), which, in the non-relativistic (NR) limit of DSR (*i.e.* $v_i \ll u_{i,\text{int}}(E)$), reduces to (cf. Eq. (5.12))

$$E_{\text{int,NR}} = m_{\text{in,int}} u_{i,\text{int}}^2(E) = m_{\text{in,int}} c^2 \frac{b_{0,\text{int}}^2(E)}{b_{i,\text{int}}^2(E)}. \tag{14.6}$$

In the case of the gravitational metric (see Eq. (8.28)), we have

$$\frac{b_{0,\text{grav}}^2(E)}{b_{3,\text{grav}}^2(E)} = 1, \quad \forall E \in R_0^+. \tag{14.7}$$

Therefore, for $i = 3$, Eqs. (14.5) and (14.6) become, respectively ($v_3 = v$):

$$E_{\text{grav}} = m_g c^2 \left[1 - \left(\frac{v}{c} \right)^2 \right]^{-1/2} = m_g c^2 \gamma \tag{14.8}$$

$$E_{\text{grav,NR}} = m_g c^2 \tag{14.9}$$

namely, the gravitational energy takes its standard, special-relativistic values.

This means that the special characterization (corresponding to the choice $i = 3$) of Eqs. (14.5) and (14.6) within the framework of DSR relates the gravitational interaction with SR, which is — as well known — based on the electromagnetic interaction in its Minkowskian form.

14.2. The Electron as a Fundamental Particle and its "Geometrical" Mass

Let us consider for E the threshold energy of the gravitational interaction:

$$E = E_{0,\text{grav}}. \tag{14.10}$$

By definition of $E_{0,\text{grav}}$, one has therefore:

$$\eta_{\mu\nu,\text{grav}}(E) = \text{diag}(1, -b_{1,\text{grav}}^2(E), -b_{2,\text{grav}}^2(E), -1)$$
$$\overset{\text{ESC off}}{=} \delta_{\mu\nu} \left[\delta_{\mu 0} - \delta_{\mu 1} b_{1,\text{grav}}^2(E) - \delta_{\mu 2} b_{2,\text{grav}}^2(E) - \delta_{\mu 3}\right],$$
$$\forall E \in (0, E_{0,\text{grav}}].$$

Notice that at the energy $E = E_{0,\text{grav}}$ the electromagnetic metric (7.8) is Minkowskian, too (because $E_{0,\text{grav}} > E_{0,\text{em}}$).

On the basis of the previous considerations, it seems reasonable to assume that the physical object (particle) p with a rest energy (*i.e.* gravitational mass) just equal to the threshold energy $E_{0,\text{grav}}$, namely

$$E_{0,\text{grav}} = m_{g.,p} c^2, \tag{14.11}$$

must play a fundamental role for either e.m. and gravitational interaction. We can e.g. hypothesize that p corresponds to the lightest mass eigenstate which experiences both force fields (*i.e.* from a quantum viewpoint, coupling to the respective interaction carriers, the photon and the graviton). As a consequence, p must be intrinsically stable, due to the impossibility of its decay in lighter mass eigenstates, even in the case such a particle is subject to weak interaction, too (*i.e.* it couples to all gauge bosons of the Glashow–Weinberg–Salam group $SU(2) \times U(1)$, not only to its electromagnetic charge sector).

Since, as above seen, for $E = E_{0,\text{grav}}$ the electromagnetic metric is Minkowskian, too, it is natural to assume, for p:

$$m_{\text{in},p,\text{e.m.}} = m_{\text{in},p} \tag{14.12}$$

namely its inertial mass is that measured with respect to the electromagnetic metric.

Then, due to the Equivalence Principle (see Eq. (14.2)), the mass of p is characterized by

$$p : \begin{cases} m_{\text{in},p,\text{grav}} = m_{g,p} \\ m_{\text{in},p,\text{e.m.}} = m_{\text{in},p}. \end{cases} \qquad (14.13)$$

Therefore, for such a fundamental particle the LLI breaking factor (14.1) of the e.m. interaction becomes:

$$\delta_{\text{e.m.}} = \frac{m_{\text{in},p} - m_{g,p}}{m_{\text{in},p}} = 1 - \frac{m_{g,p}}{m_{\text{in},p}} \Leftrightarrow m_{g,p} = m_{\text{in},p} \left(1 - \delta_{\text{e.m.}}\right). \qquad (14.14)$$

Replacing (14.11) in (14.14) yields:

$$E_{0,\text{grav}} = m_{in,p} \left(1 - \delta_{\text{e.m.}}\right) c^2 \Leftrightarrow m_{in,p} = \frac{E_{0,\text{grav}}}{c^2} \frac{1}{1 - \delta_{\text{e.m.}}}. \qquad (14.15)$$

Equation (14.15) allows one to evaluate the inertial mass of p from the knowledge of the electromagnetic LLI breaking parameter $\delta_{\text{e.m.}}$ and of the threshold energy $E_{0,\text{grav}}$ of the gravitational metric.

The lowest limit to the LLI breaking factor of electromagnetic interaction has been determined by the experiment (based on the detection of a DC voltage across a conductor induced by the steady magnetic field of a coil), discussed in Chap. 13. On account of the definition (13.14) of the isotropic LLI breaking parameter, and the definition (14.1), the value found corresponds to

$$1 - \delta_{\text{e.m.}} \cong 4 \cdot 10^{-11}. \qquad (14.16)$$

Then, inserting the value (8.24) for $E_{0,\text{grav}}$ and (14.16) in (14.15), one gets

$$m_{\text{in},p} = \frac{E_{0,\text{grav}}}{c^2} \frac{1}{1 - \delta_{\text{e.m.}}} \geq \frac{2 \cdot 10^{-5}}{4 \cdot 10^{-11}} \frac{\text{eV}}{c^2} = 0.5 \frac{\text{MeV}}{c^2} = m_{\text{in},e} \qquad (14.17)$$

(with $m_{\text{in},e}$ being the electron mass) where the $\geq$ is due to the fact that in general the LLI breaking factor constitutes an *upper limit* (*i.e.* it sets the scale *under which* a violation of LLI is expected). If the coil experiment does indeed provide evidence for a LLI breakdown (as it seems the case, although further confirmation is needed), Eq. (14.17) yields $m_{\text{in},p} = m_{\text{in},e}$. We find therefore the amazing result that *the fundamental particle p is nothing but the electron e^- (or its antiparticle e^+).*[b] The electron is indeed the lightest

---

[b]Of course, this last statement does strictly holds only if the CPT theorem mantains its validity in the DSR framework, too. Although this problem has not yet been addressed in general on a formal basis, it can be stated that it holds true in the case considered, since the energy value $E = E_{0,\text{grav}}$ corresponds to the minkowskian form of both electromagnetic and gravitational metric.

massive lepton (pointlike, non-composite particle) with electric charge, and therefore subjected to gravitational, electromagnetic and weak interactions, but unable to weakly decay due to its small mass. Consequently, e^- (e^+) shares all the properties we required for the particle p, whereby it plays a fundamental role for gravitational and electromagnetic interactions.

It was therefore shown that within DSR it is possible — on the basis of simple and plausible assumptions — to evaluate the inertial mass of the electron e^- (and therefore of its antiparticle, the positron e^+) by exploiting the expression of the relativistic energy in the deformed Minkowski space $\tilde{M}_4(E)_{E \in R_0^+}$, the explicit form of the phenomenological metric describing the gravitational interaction (in particular its threshold energy), and the LLI breaking parameter for the electromagnetic interaction $\delta_{\text{e.m.}}$. Let us stress that no use has been made of the electric charge e of the electron.

Therefore, the inertial properties of one of the fundamental constituents of matter and of Universe do find a "geometrical" interpretation in the context of DSR, by admitting for local violations of standard Lorentz invariance.

CHAPTER 15

TOWARD THE FIFTH DIMENSION

15.1. From LLI Breakdown to Energy as Fifth Dimension

Both the analysis of the physical processes considered in deriving the phenomenological energy-dependent metrics for the four fundamental interactions, and the coil experiment discussed in Chap. 13, seem to provide evidence (indirect and direct, respectively) for a breakdown of LLI invariance. But it is well known that, in general, the breakdown of a symmetry is the signature of the need for a *wider*, *exact* symmetry. In the case of the breaking of a space-time symmetry — as the Lorentz one — this is often related to the possible occurrence of higher-dimensional schemes. It will be shown that this is indeed the case, and that energy does in fact represent an extra dimension.

In the description of interactions by energy-dependent metrics, we see that energy plays in fact a *dual* role. On one side, as more and more stressed, it constitutes a dynamical variable. On the other hand, it represents a parameter characteristic of the phenomenon considered (and therefore, for a given process, it cannot be changed at will). In other words, when describing a given process, the deformed geometry of space-time (in the interaction region where the process is occurring) is "frozen" at the situation described by those values of the metric coefficients $\left\{b_\mu^2(E)\right\}_{\mu=0,1,2,3}$ corresponding to the energy value of the process considered. Namely, a fixed value of E determines the space-time structure of the interaction region at that given energy. In this respect, therefore, the energy of the process is to be considered as *a geometrical quantity* intimately related to the very geometrical structure of the physical world. In other words, from a geometrical point of view, all goes on as if were actually working on "slices" (sections) of a five-dimensional space, in which the extra dimension is just represented by the energy.

The simplest way to take account of (and to make explicit) the double role of energy in DSR is assuming that E represents a fifth metric dimension — on the same footing of space and time — and therefore to embed the 4-d deformed Minkowski space $\tilde{M}_4(E)$ of DSR in a 5-d (Riemann) space $\mathcal{R}_5$. This leads to build up a "Kaluza–Klein-like" scheme, with energy as fifth dimension, we shall refer to in the following as *Five-Dimensional Deformed Special Relativity* (DSR5).[84,85]

Let's recall that the use of momentum components as metric variables on the same foot of the space-time ones can be traced back to Ingraham.[86] On the contrary, it was just shown by T. D. Lee that time (namely, a space-time coordinate) can be used as a dynamical variable. Moreover, many authors (starting from Dirac[87]) treated mass as a dynamical variable in the context of scale-invariant theories of gravity.[88,89] Such a point of view has been advocated also in the framework of modern Kaluza–Klein theories by the so-called "Space-Time-Mass" (STM) theory, in which the fifth dimension is the rest mass, proposed by Wesson and studied in detail by a number of authors.[90]

The formalism of DSR5 is presently still being developed, and we shall confine ourselves to sketch only its main features in the next sections.

15.2. Embedding of Deformed Minkowski Space in a Five-Dimensional Riemannian Space

We assume then a 5-dimensional space-time $\mathcal{R}_5$ endowed with the energy-dependent metric[a]:

$$\eta_{AB}^{(5)}(E) \equiv \text{diag}(b_0^2(E), -b_1^2(E), -b_2^2(E), -b_3^2(E), f(E)) \stackrel{\text{ESC off}}{=}$$
$$= \delta_{AB} \left(b_0^2(E)\delta_{A0} - b_1^2(E)\delta_{A1} - b_2^2(E)\delta_{A2} - b_3^2(E)\delta_{A3} + f(E)\delta_{A5} \right).$$
$$(15.1)$$

It follows from Eq. (15.1) that E, which is an independent *non-metric* variable in DSR, becomes a *metric* coordinate in $\mathcal{R}_5$. Then, whereas $\eta_{\mu\nu}(E)$ (given by (3.3)) is a deformed, Minkowskian metric tensor, $\eta_{AB}^{(5)}(E)$ is a genuine Riemannian metric tensor.

Therefore, the infinitesimal interval of $\mathcal{R}_5$ is given by:

$$ds_{(5)}^2 \equiv \eta_{AB}^{(5)}(E)dx^A dx^B$$
$$= b_0^2(E) \left(dx^0 \right)^2 - b_1^2(E) \left(dx^1 \right)^2 - b_2^2(E) \left(dx^2 \right)^2$$
$$- b_3^2(E) \left(dx^3 \right)^2 + f(E) \left(dx^5 \right)^2$$

---

[a]In the following, capital Latin indices take values in the range $\{0, 1, 2, 3, 5\}$, with index 5 labelling the fifth dimension.

$$= b_0^2(E)c^2 (dt)^2 - b_1^2(E) (dx^1)^2 - b_2^2(E) (dx^2)^2$$
$$- b_3^2(E) (dx^3)^2 + f(E)l_0^2 (dE)^2 \tag{15.2}$$

where we have put

$$x^5 \equiv l_0 E, \quad l_0 > 0. \tag{15.3}$$

The constant l_0 provides the dimensional conversion energy $\rightarrow$ length (it has therefore the dimensions of a force).

The coefficients $\{b_\mu^2(E)\}$ of the metric of $\tilde{M}_4(E)$ can be therefore expressed as

$$\left\{ b_\mu \left(\frac{E}{E_0} \right) \right\} \equiv \left\{ b_\mu \left(\frac{x^5}{x_0^5} \right) \right\} = \{b_\mu(x^5)\}, \quad \forall \mu = 0, 1, 2, 3 \tag{15.4}$$

where we put

$$x_0^5 \equiv l_0 E_0. \tag{15.5}$$

As to the fifth metric coefficient, one assumes that it too is a function of the energy only: $f = f(E) \equiv f(x^5)$. (although, in principle, nothing prevents from assuming that, in general, f may depend also on space-time coordinates $\{x^\mu\}$, $:f = f(\{x^\mu\}, x^5)$). Unlike the other metric coefficients, it may be $f(E) \lessgtr 0$. Therefore, *a priori*, the energy dimension may have either a timelike or a spacelike signature in $\mathcal{R}_5$, depending on $\mathrm{sgn}(f(E)) = \pm 1$.

We recall that in general, in the framework of 5-d Kaluza–Klein (KK) theories, the fifth dimension must be necessarily spacelike, since, in order to avoid the occurrence of causal (loop) anomalies, the number of timelike dimensions cannot be greater than one.[91] But it is worth to stress that the present theory is not a Kaluza–Klein one. In fact in the framework of DSR5 the cylindricity condition — a basic feature of "true" KK theories — is not implemented, and even reversed. Actually, the metric tensor $\eta_{AB}^{(5)}(E)$ depends only on the fifth coordinate E. As a consequence, one does not implement the compactification of the extra coordinate (one of the main methods of implementing the cylindricity condition in modern iperdimensional KK theories), which remains therefore extended (*i.e.* with infinite compactification radius). The problem of the possible occurrence of causal anomalies in presence of more timelike dimensions is then left open in the "pseudo-Kaluza–Klein" context of DSR5. There is therefore an uncertainty in the sign of the energy metric coefficient $f(E)$. In particular, it cannot be excluded *a priori* that the signature of E can change. This occurs whenever the function $f(E)$ does vanish for some energy values. As a consequence, in correspondence to the energy values which are zeros of $f(E)$, the metric $\eta_{AB}^{(5)}(E)$ is *degenerate*.

DSR5 is therefore a Kaluza–Klein-like scheme, whereby now the extra parameter is a physically sensible dimension. Thus, on this respect, such a formalism belongs to the class of noncompactified KK theories.[92] Moreover, it has some connection with Wesson's STM theory.[90] Both in the DSR5 formalism and in the STM theory (at least in its more recent developments) it is assumed that all metric coefficients do in general depend on the fifth coordinate. Such a feature distinguishes either models from true Kaluza–Klein theories. However, the present approach differs from the STM model — as well as from similar ones, like *e.g.* the Space-Time-Mass-(Electric) Charge (S.T.M.C.) by Fukui[93] — at least in the following main respects:

(i) Its physical motivations are based on the phenomenological analysis of Part III, and therefore are not merely speculative;

(ii) The fact of assuming *energy* (which is a true variable), and not rest mass (which instead is an invariant), as fifth dimension[b];

(iii) the *local* (and not *global*) nature of the five-dimensional space $\mathcal{R}_5$, whereby the energy-dependent deformation of the four-dimensional space-time is assumed to provide a geometrical description of the interactions.

15.3. Einstein's Equations in Vacuum

The (vacuum) Einstein equations in the space $\Re_5$ are

$$R_{AB} - \frac{1}{2}\eta_{AB}^{(5)}R = \Lambda\eta_{AB}^{(5)} \tag{15.6}$$

where R_{AB} and $R = R_A^A$ are the five-dimensional Ricci tensor and scalar curvature, respectively, while Λ is the "cosmological" constant. Λ is assumed to be a genuine constant, although it might also, in principle, depend on both the energy E and the space-time coordinates x: $\Lambda = \Lambda(x, E)$.

In the case of spatial isotropy $(b_1(E) = b_2(E) = b_3(E) = b(E))$, these equations can be reduced to the following form[84,85]:

$$\begin{cases} 3(-2b''f + b'f') = 4\Lambda bf^2; \\ f\left[b_0^2(b')^2 - 2b_0 b_0' bb' - 4b_0^2 bb'' - 2b_0 b_0'' b^2 + b^2(b_0')^2\right] \\ \quad + b_0 bf'(2b_0 b' + b_0' b) = 4\Lambda b_0^2 b^2 f^2; \\ 3b'(b_0 b)' = -4\Lambda b_0 b^2 f. \end{cases} \tag{15.7}$$

---

[b]In this respect, therefore, the DSR5 formalism rensembles more the one due to Ingraham.[86]

where a prime denotes derivation with respect to E and we put $c = \ell_0 = 1$. Among the solutions corresponding to $\Lambda = 0$, one finds[c]:

$$\begin{cases} f(E) = \text{const.}, \\ b_0(E) = \left(1 + \dfrac{E}{E_0}\right)^2, \\ b(E) = \text{const.}, \end{cases} \tag{15.8}$$

where E_0 is a constant having the dimensions of energy. This solution is in fact the unique solution corresponding to a constant coefficient $f(E)$ in a larger family in which $b_0(E)$ is arbitrary and (f, b_0) satisfy the ordinary differential equation:

$$-2b_0 b_0'' f + (b_0')^2 f + b_0 b_0' f' = 0. \tag{15.9}$$

Moreover, the metric (15.8) coincides with the phenomenological gravitational metric (8.27) in the hypothesis of spatial isotropy.

Other interesting classes of solutions to Eqs. (15.7) for $\Lambda = 0$ can be obtained by the ansatz that all the metric coefficients in $\eta_{\mu\nu}(E)$(Eq. (3.3)) are pure powers of E, *i.e.*

$$\begin{cases} b_0(E) = (E/E_0)^q; \\ b_1(E) = (E/E_0)^m; \\ b_2(E) = (E/E_0)^n; \\ b_3(E) = (E/E_0)^p, \end{cases} \tag{15.10}$$

while for the dimensional parameter $f(E)$ one assumes simply

$$f(E) = E^r. \tag{15.11}$$

In this case, the Einstein equations reduce to the algebraic system:

$$\begin{cases} (2+r)(p+m+n) - m^2 - n^2 - p^2 - mn - mp - np = 4\Lambda E^{r+2}; \\ (2+r)(p+q+n) - n^2 - p^2 - q^2 - np - nq - pq = 4\Lambda E^{r+2}; \\ (2+r)(p+q+m) - m^2 - p^2 - q^2 - mp - mq - pq = 4\Lambda E^{r+2}; \\ (2+r)(q+m+n) - m^2 - n^2 - q^2 - mn - mq - nq = 4\Lambda E^{r+2}; \\ \qquad\qquad mn + mp + mq + np + nq + pq = -4\Lambda E^{r+2}, \end{cases} \tag{15.12}$$

which admit (at least) twelve classes of solutions. They include as special cases *all* the phenomenological metrics discussed in Part III. In particular,

---

[c]Since Λ is related to the vacuum energy in General Relativity and experimental evidence shows that $\Lambda \simeq 3 \cdot 10^{-52} \text{m}^{-2}$, it is possible to assume $\Lambda = 0$ as far as one is not interested in quantum effects.

the metric (15.8) can be obtained as the only metric which lives in the intersection of three of the relevant classes, namely it is obtained by setting:

$$q = 2, \quad m = n = p = r = 0, \tag{15.13}$$

which obviously reduces to (8.27) by a rescaling and a translation of the energy parameter E_0.

Needless to say, the fact that it is possible — by the mere consideration of the Einstein equations in vacuum — to recover all the phenomenological metrics for the four fundamental interactions, is a further confirmation of the fact that DSR is indeed naturally embedded in the five-dimensional scheme of DSR5.

References

1. E. Recami and R. Mignani: *Riv. Nuovo Cimento* **4**, n. 2 (1974), and references therein.
2. R. Penrose: *The Emperor's new mind* (Oxford University Press, 1989).
3. F. Cardone and R. Mignani: *"On a nonlocal relativistic kinematics"*, INFN preprint n.910 (Roma, Nov. 1992).
4. F. Cardone and R. Mignani: *Found. Phys.* **29**, 1735 (1999).
5. F. Cardone and R. Mignani: *Ann. Fond. L. de Broglie* **25**, 165 (2000).
6. For a review of Finsler's generalization of Riemannian spaces, see *e.g.* M. Matsumoto: *Foundation of Finsler Geometry and Special Finsler Spaces* (Kaiseisha Otsu, 1986), and references therein.
7. For a review, see G. Yu. Bogoslovsky: *Fortsch. Phys.* **42**, 2 (1994).
8. For a review of Lie-isotopic theories, see R. M. Santilli: *Found. Phys.* **27**, 625 (1997).
9. H. Nielsen and I. Picek: *Nucl. Phys.* **B211**, 269 (1983).
10. See *e.g.* P. Kosiński and P. Maślanka, in *From Field Theory to Quantum Groups*, eds. B. Jancewicz, J. Sobczyk (World Scientific, Singapore, 1996).
11. W. Drechsler: in W. Drechsler and M. E. Mayer, *Fiber Bundle Techniques in Gauge Theories* (Springer-Verlag, Berlin, 1977), pg. 145.
12. S. Coleman and S. L. Glashow: *Phys. Lett.* **B405**, 249 (1997).
13. F. Cardone and R. Mignani: *Ann. Fond. L. de Broglie* **23**, 173 (1998).
14. F. Cardone, R. Mignani and V. S. Olkhovski: *J. de Phys. I (France)* **7**, 1211 (1997).
15. F. Cardone, R. Mignani, and V. S. Olkhovski: *Modern Phys. Lett.* **B14**, 109 (2000).
16. Ph. Balcou and L. Dutriaux: *Phys. Rev. Lett.* **78**, 851 (1997).
17. G. Barton and K. Scharnhorst: *J. Phys.* **A26**, 2037 (1993), and references therein.
18. Such a result can be traced back to T. Levi-Civita: *"The Absolute Differential Calculus"* (Blackie & Son, 1954), p. 403.
19. I. T. Drummond and S. J. Hathrell: *Phys. Rev.* **D2**, 345 (1980).
20. E. H. Hauge and J. A. Støvneng: *Rev. Mod. Phys.* **61**, 917 (1989).
21. V. S. Olkhovsky and E. Recami: *Phys. Rep.* **214**, 339 (1992).
22. R. Landauer and Th. Martin: *Rev. Mod. Phys.* **66**, 217 (1994).
23. T. E. Hartman: *J. Appl. Phys.* **33**, 3427 (1962).
24. J. R. Fletcher: *J. Phys.* **C18**, L55 (1985).

25. S. Bosanac: *Phys. Rev.* **A28**, 577 (1983).
26. Th. Martin and R. Landauer: *Phys. Rev.* **A45**, 2611 (1992).
27. R. Y. Chiao, P. G. Kwiat and A. M. Steinberg: *Physica* **B175**, 257 (1991).
28. A. Ranfagni, D. Mugnai, P. Fabeni and G. P. Pazzi: *Appl. Phys. Lett.* **58**, 774 (1991).
29. A. Enders and G. Nimtz: *J. Phys. I (France)* **2**, 1693 (1992).
30. A. Enders and G. Nimtz: *J. Phys. I (France)* **3**, 1089 (1993).
31. A. Enders and G. Nimtz: *Phys. Rev.* **E48**, 632 (1993).
32. G. Nimtz, A. Enders and H. Spieker: *J. Phys. I (France)* **4**, 1 (1994).
33. W. Heitmann and G. Nimtz: *Phys. Lett.* **A196**, 154 (1994).
34. A. Enders and G. Nimtz: *Phys. Rev.* **B47**, 9605 (1993).
35. G. Nimtz, A. Enders and H. Spieker: *J. Phys. I (France)* **4**, 565 (1994).
36. A. Ranfagni, P. Fabeni, G. P. Pazzi and D. Mugnai: *Phys. Rev.* **E48**, 1453 (1993).
37. D. Mugnai, A. Ranfagni and L. S. Schulman: *Phys. Rev.* **E55**, 3593 (1997).
38. D. Mugnai, A. Ranfagni and L. Ronchi: *Phys. Lett.* **A247**, 281 (1998).
39. A. M. Steinberg, P. G. Kwait and R. Y. Chiao: *Phys. Rev. Lett.* **71**, 708 (1993).
40. Ch. Spielmann, R. Szipocs, A. Singl and F. Krausz: *Phys. Rev. Lett.* **73**, 2308 (1994).
41. V. Laude and P. Tournois: *J. Opt. Soc. Am.* **B16**, 194 (1999).
42. F. Cardone and R. Mignani: *Phys. Lett.* **A306**, 265 (2003).
43. H. P. Friis: *IEEE Spectrum* **8**, 55 (1971).
44. A. Ranfagni and D. Mugnai: *Phys. Rev.* **E58**, 6742 (1998).
45. C. O. Alley: "Relativity and Clocks", in *Proc. of the 33rd Annual Symposium on Frequency Control,* Elec. Ind. Ass., Washington, D.C. (1979).
46. See C. M. Will: *Theory and Experiment in Gravitational Physics* (Cambridge Univ. Press, rev. ed. 1993), and references therein.
47. F. Cardone and R. Mignani: *Int. J. Modern Phys.* **A14**, 3799 (1999).
48. T. Van Flandern and J.-P. Vigier: *Found. Phys.* **32**, 1031 (2002).
49. T. Van Flandern: *Phys. Lett.* **A250**, 1 (1998).
50. W. D. Walker and J. Dual: in *Gravitational Waves — Proc. of the Second Edoardo Amaldi Conference,* eds. E. Coccia, G. Pizzella and G. Veneziano (World Scientific, Singapore, 1997).
51. See e.g. L. Rosenfeld, in *Proc. Fourteenth Solvay Conf.* (Wiley-Interscience, N.Y., U.S.A., 1968), p. 232.
52. For Bohr's reply to Einstein's criticism see N. Bohr, in *Albert Einstein: Philosopher-Scientist,* ed. P. Shilpp (Tudor, N.Y., U.S.A., 1949), p. 199.
53. See A. Pais: *"Subtle is the Lord ..." — The Science and the Life of Albert Einstein* (Oxford Univ. Press, 1982).
54. G. Amelino-Camelia, J. Ellis, N. E. Mavromatos, D. V. Nanopoulos and S. Sarkar: *Nature* **393**, 763 (1998).
55. P. Kaaret: *Astron. Astrophys.* **345**, L32 (1999).
56. P. A. M. Dirac: *Nature* **139**, 323 (1937).
57. M. J. Drinkwater, J. K. Webb, J. Barrow and V. V. Flambaum: *Mon. Not. Roy. Astron. Soc.* **295**, 457 (1998).

58. See *e.g.* A. V. Ivanchik, A. Y. Potekhin and D. A. Varshalovich: *Astron. Astrophys.* (1998), and references therein.
59. See *e.g.* M. B. Green and J. H. Schwarz: *Superstring Theory* (Cambridge Univ. Press, 1987).
60. P. Sisterna and H. Vucetich: *Phys. Rev.* **D41**, 1034 (1990).
61. S. H. Aronson, G. J. Bock, H.-Y. Chang and E. Fishbach: *Phys. Rev. Lett.* **48**, 1306 (1982); *Phys. Rev.* **D28**, 495 (1983).
62. N. Grossman *et al.*: *Phys. Rev. Lett.* **59**, 18 (1987).
63. F. Cardone, R. Mignani and R. M. Santilli: *J. Phys.* **G18**, L61, L141 (1992).
64. See *e.g.* B. Lörstad: *Correlations and Multiparticle Production (CAMP)*, eds. M. Pluenner, S. Raha and R. M. Weiner (World Scientific, Singapore, 1991), and references therein.
65. M. Gaspero: *Sov. J. Nucl. Phys.* **55**, 795 (1992); *Nucl. Phys.* **A562**, 407 (1993); *ibidem*, **588**, 861 (1995).
66. M. Gaspero and A. De Pascale: *Phys. Lett.* **B358**, 146 (1995).
67. CPLEAR collaboration, R. Adler *et al.*: *Nucl. Phys.* **A558**, 43c (1993); *Z. Phys.* **C63**, 541 (1994).
68. F. Cardone and R. Mignani: *JETP* **83**, 435 [*Zh. Eksp. Teor. Fiz.* **110**, 793] (1996).
69. F. Cardone, M. Gaspero and R. Mignani: *Eur. Phys. J.* **C4**, 705 (1998).
70. UA1 Collaboration: *Phys. Lett.* **B226**, 410 (1989).
71. L. Kostro and B. Lange: *Phys. Essays* **10** (1999), and references therein.
72. R. Jackiw: "Chern–Simons violation of Lorentz and PCT symmetries in electrodynamics" (hep-ph/9811322, 13 Nov. 1998), and references therein.
73. A. Kostelecky, ed.: *CPT and Lorentz Symmetry*, I and II (World Scientific, Singapore, 1999 and 2002); and references therein.
74. U. Bartocci and M. Mamone Capria: *Found. Phys.* **21**, 787 (1991).
75. U. Bartocci, F. Cardone and R. Mignani: *Found. Phys. Lett.* **14**, 51 (2001).
76. F. Cardone and R. Mignani: *Physics Essays* **13**, 643 (2000).
77. F. Cardone and R. Mignani: "On possible experimental evidence for a breakdown of local Lorentz invariance", in *Gravitation, Electromagnetism and Cosmology: Toward a New Synthesis* (Proc. Int. Conf. On Redshifts and Gravitation in a Relativistic Universe, Cesena, Italy, Sept. 17–20, 1999), ed. by K. Rudnicki (Apeiron, Montreal, 2001), p. 165.
78. F. Cardone and R. Mignani: "On the limit of validity of Lorentz invariance", Proc. Int. Conf. *"Physical Interpretations of Relativity Theory VII"*, (Imperial College, London, 15–18 Sept. 2000), *Late Papers & Supplementary Papers*, M. Duffy ed., p. 40.
79. O. D. Jefimenko: *Electricity and Magnetism,* 2nd ed. (Electret Sci., Star City, 1989).
80. M. A. Heald: *Amer. J. Phys.* **52**, 522 (1984).
81. A. K. T. Assis, W. A. Rodrigues, Jr. and A. J. Mania: *Found. Phys.* **29**, 729 (1999).
82. See *e.g.* L. Kostro: *Einstein and the Aether* (Apeiron, Montreal, 2000).
83. F. Cardone, A. Marrani and R. Mignani: *Electromagnetic Phenomena* (Special Issue for Dirac's Centenary) **3**, 11 (2003).

84. F. Cardone, M. Francaviglia and R. Mignani: *Gen. Rel. Grav.* **30**, 1619 (1998); *ibidem*, **31**, 1049 (1999).
85. F. Cardone, M. Francaviglia and R. Mignani: *Found. Phys. Lett.* **12**, 281, 347 (1999).
86. R. L. Ingraham: Nuovo Cim. **9**, 87 (1952).
87. P. A. M. Dirac: *Proc. R. Soc. (London)* **A333**, 403 (1973); *ibidem*, **A338**, 439 (1974).
88. F. Hoyle and J. V. Narlikar: *Action at a Distance in Physics and Cosmology* (Freeman, N.Y., 1974).
89. V. Canuto, P. J. Adams, S.-H. Hsieh and E. Tsiang: *Phys. Rev.* **D16**, 1643 (1977); V. Canuto, S.-H. Hsieh and P. J. Adams: *Phys. Rev. Lett.* **39**, 429 (1977).
90. P. S. Wesson: *Space-Time-Matter — Modern Kaluza–Klein Theory* (World Scientific, Singapore, 1999), and references therein.
91. See *e.g.* L. Sklar: *Space, Time, and Spacetime* (University of California Press, Berkeley, 1976), Chap. IV.
92. See *e.g.* T. Appelquist, A. Chodos and P. G. O. Freund (editors): *Modern Kaluza-Klein Theories* (Addison-Wesley, Menlo Park, California, 1987), and references therein.
93. T. Fukui: *Gen. Rel. Grav.* **20**, 1037 (1988); *ibidem*, **24**, 389 (1992).

Index

aberration law
  deformed, 39
Abraham, M., 75
absolute reference frame, 118
aether
  and LLI breakdown, 123
  as static gravitational field, 123
  as vacuum, 123
  drift, 123
  Einstein's conception of, 123
Alley, C. O., 67
asymptotic freedom
  geometric view to, 90

background
  gravitational radiation, 119
  thermal radiation, 71, 118
Bartocci experiment, 109
Bartocci, U., 107, 108, 123
Berkeley experiment, 58
Bogoslowski, G. Yu., 7, 8
Bohr, N., 73, 74
boost
  deformed, 29
      derivation of, 35
      in a generic direction, 28, 30
      symmetrization of, 30
Bose-Einstein correlation, 14, 37, 81
  canonical treatment of, 81
  deformed function, 83
      anisotropic, 84
      isotropic, 84
  DSR treatment of, 82
  fireball, 83
  function, 81
  incoherence parameter, 82

Cartesian frame, 48
Cauchy stress tensor, 48
chaoticity, 82
cilindricity condition, 133
Clausius postulate, 107
clock rate, 74
  in a gravitational field, 14
  in gravitational field, 67
coherence wavelength, 59
Cohn, E., 123
Coleman, S., 10, 105, 120
Cologne experiment, 55
  and Friis law, 63
  high-pass filter behavior, 61
confinement
  geometric view to, 91
cosmological constant
  in five dimensions, 134
CP invariance, 77
CPT symmetry, 105
critical
  angle, 46
  length, 64
  wavelength, 46
cutoff
  frequency, 54, 55, 61–63, 65
      waveguide, 43
  length, 64, 65
      waveguide, 46, 61

D'Alembert
  operator
      deformed, 41
De Broglie mass, 59

141

Printed in the United States
By Bookmasters